Learning Resource Manual to Accompany

HUMAN PHYSIOLOGY

FROM CELLS TO SYSTEMS

Lauralee Sherwood
Department of Physiology
School of Medicine
West Virginia University

West Publishing Company
St. Paul New York Los Angeles San Francisco

50 W. Kellogg Boulevard
P.O. Box 64526
St. Paul, MN 55164-1003

Printed in the United States of America
96 95 94 93 92 91 90 89 8 7 6 5 4 3 2 1 0
ISBN 0-314-52508-4

CONTENTS

C H A P T E R 1

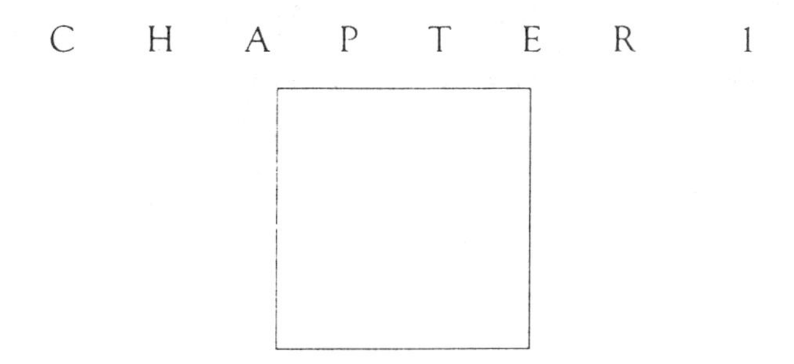

HOMEOSTASIS: THE FOUNDATION OF PHYSIOLOGY

Introduction (text page 2)

Levels of Organization in the Body (text page 3)

Contents

Cells are the basic units of life. p. 3

Cells are progressively organized into tissues, organs, systems, and, finally, the whole body. p. 3

Section Synopsis

The human body is composed of an interactive society of cells, which are the basic units of both structure and function. Each cell, whether it exists as a solitary unicellular organism or as part of a complex multicellular organism, performs certain basic functions essential for its own survival. Among these basic cell functions are obtaining O_2 and nutrients, which the cell uses to acquire energy; eliminating wastes; synthesizing needed cellular components; being sensitive to changes in the surrounding environment; controlling movement of materials within the cell and between the cell and its environment; and reproducing.

In multicellular organisms, each cell performs in addition to these fundamental cell functions a specialized activity, which is usually an elaboration of one of the basic cell functions. The body's cells are highly organized into functional groupings. Cells of similar structure and specialized activity are organized into tissues, of which there are four primary types: (1) muscle tissue, which is specialized for contraction and force generation; (2) nervous tissue, which is specialized for initiation and transmission of electrical impulses; (3) epithelial tissue, which lines and covers various body surfaces and cavities and which also forms secretory glands; and (4) connective tissue, which connects, supports, and anchors various body parts.

Tissues are organized into organs, which are structures composed of several types of primary tissue organized to perform a particular function or functions. Organs are further organized into body systems, which are collections of organs that perform related functions and interact to accomplish a common activity essential for survival of the whole body. Organ systems, in turn, compose the whole body.

Learning Check (Answers on p. A-1)

A. Multiple Choice

1. Which of the following activities is <u>not</u> carried out by every cell in the body?

   a. obtaining O_2 and nutrients

   b. performing chemical reactions to acquire energy for the cell's use

   c. eliminating wastes

   d. controlling to a large extent exchange of materials between a cell and its external environment

   e. reproducing

2. Which of the following is the proper progression of the levels of organization in the body?

   a. cells, organs, tissues, body systems, whole body

   b. cells, tissues, organs, body systems, whole body

   c. cells, tissues, organs, whole body, body systems

   d. cells, organs, tissues, whole body, body systems

   e. cells, tissues, body systems, organs, whole body

3. Which of the following is <u>not</u> a type of connective tissue?

   a. bone

   b. blood

   c. the spinal cord

   d. tendons

   e. the tissue that attaches epithelial tissue to underlying structures

B. True/False

T/F 1. A mechanistic explanation of why a person sweats is to cool off.

T/F 2. The term <u>tissue</u> can apply to either one of the four primary tissue types or to a particular organ's aggregate of cellular and extracellular components.

T/F 3. Glands are formed during embryonic development by pockets of epithelial tissue that dip inward from the surface.

T/F 4. Cells in a multicellular organism have specialized to such a great extent that they share little in common with single-celled organisms.

T/F 5. Cellular specializations are usually a modification or elaboration of one of the basic cell functions.

C. Fill-in-the-blank

1. The smallest unit capable of carrying out the processes associated with life is the ________________.

2. The four primary types of tissue are ________________, ________________, ________________, and ________________.

3. ________________ refers to the release from a cell, in response to appropriate stimulation, of specific products that have in large part been synthesized by the cell.

4. ________________ glands secrete through ducts to the outside of the body, whereas ________________ glands release their secretory products, known as ________________, internally into the blood.

Concept of Homeostasis **(text page 5)**

Contents

Body cells are in contact with a privately maintained internal environment instead of with the external environment that surrounds the body. p. 5

Each cell requires homeostasis, and each cell, as part of an organized system, contributes to homeostasis. p. 6

A Closer Look at Exercise Physiology - What is Exercise Physiology? p. 7

Negative feedback is a common regulatory mechanism for maintaining homeostasis. p. 11

Disruptions in homeostasis can lead to illness and death. p. 12

Section Synopsis

Homeostasis refers to maintenance of a dynamic steady state within the internal fluid environment that bathes all of the body's cells. Because the body's cells are not in direct contact with the external environment, cell survival depends on maintenance of a stable internal fluid environment with which the cells directly make exchanges. For example, O_2 and nutrients must constantly be replenished in the internal environment to keep pace with the rate at which the cells use these materials for energy production. The factors of the internal environment that must be homeostatically maintained are: (1) its concentration of nutrient molecules; (2) its concentration of O_2 and CO_2; (3) its concentration of waste products; (4) its pH; (5) its concentration of salt and other electrolytes; (6) its temperature; and (7) its volume and pressure.

The functions performed by each of the eleven body systems are aimed at maintaining homeostasis. The body systems' functions ultimately depend on the specialized activities of the cells composing each system. Thus, each cell requires homeostasis and each cell contributes to homeostasis.

The two general categories of control systems that regulate the body systems' various activities to maintain homeostasis are: (1) intrinsic controls, which are inherent compensatory responses of an organ to a change; and (2) extrinsic controls, which are regulatory responses of organs induced by the nervous and endocrine systems. Both types of control systems generally operate on the principle of negative feedback. With negative feedback, a change in a regulated variable triggers a response that drives the variable in the opposite direction of the initial change, thus opposing the change.

Pathophysiological states ensue when one or more of the body systems fails to function properly so that an optimal internal environment can no longer be maintained. Serious homeostatic disruption leads to death.

Learning Check (Answers on p. A-2)

A. Fill-in-the-blank

1. The body cells are in direct contact with and make life-sustaining exchanges with the ________________.

2. The internal environment consists of the ______________, which is made up of ____________, the fluid portion of the blood, and ________________, which surrounds and bathes all cells.

3. ____________ controls are inherent to an organ, whereas ____________ controls are regulatory mechanisms initiated outside of an organ that alter the activity of the organ.

4. ________________ exists when a change in a regulated variable triggers a response that opposes the change.

5. _______________ refers to the abnormal functioning of the body associated with disease.

B. List the factors that must be homeostatically maintained.

C. True/False

T/F 1. Cells require homeostasis, body systems maintain homeostasis, and cells make up body systems.

T/F 2. With positive feedback, a control system's input and output continue to enhance each other.

T/F 3. To sustain life, the internal environment must be maintained in an absolutely unchanging state.

D. Matching

____ 1. circulatory system

____ 2. digestive system

____ 3. respiratory system

____ 4. urinary system

____ 5. muscular and skeletal systems

____ 6. integumentary system

____ 7. immune system

____ 8. nervous system

____ 9. endocrine system

____10. reproductive system

a. obtains O_2 and eliminates CO_2

b. support and protect body parts and allow movement

c. controls, via hormones it secretes, processes that require duration

d. transport system

e. removes wastes and excess water, salt, and other electrolytes

f. essential for perpetuation of species

g. obtains nutrients, water, and electrolytes

h. defends against foreign invaders and cancer

i. acts through electrical signals to control body's rapid responses

j. serves as protective barrier between body and external environment

Chapter in Perspective **(text page 13)**

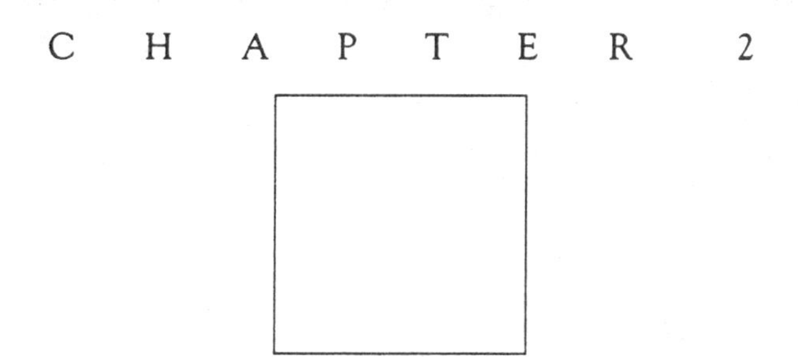

Cellular Structure and Functions

Introduction (text page 14)

Contents

Section Synopsis

The complex organization and interaction of the various chemicals within a cell confers the unique characteristics of life. Cells, in turn, are the living building blocks of the body.

Body cells, which are too small to be seen by the unaided eye, have been shown by microscopic techniques to consist of three major subdivisions: (1) the plasma membrane, which encloses the cell and separates the intracellular and extracellular fluid; (2) the nucleus, which contains deoxyribonucleic acid (DNA), the cell's genetic material; and (3) the cytoplasm, the portion of the cell's interior not occupied by the nucleus. The cytoplasm consists of organelles, which are highly organized, membrane-bound structures, dispersed within the cytosol, a complex gelatin-like mass.

Compartmentalization of specific sets of chemicals within the organelles offers two advantages. First, it permits a local concentration of key chemical factors necessary for rapid rates of important chemical reactions. Second, it permits chemical activities that would not be compatible with each other to occur simultaneously within separate organelle compartments.

Learning Check (Answers on p. A-3)

A. True/False

T/F 1. "Life" is due to the complex organization and interaction of chemical molecules within the cells.

T/F 2. The largest of the cells in the human body can be seen by the unaided eye.

T/F 3. The cells of the human body are surrounded by a cell wall.

B. Fill-in-the-blank

1. The barrier that separates and controls movement between the cellular contents and extracellular fluid is the ________________.

2. The chemical that directs protein synthesis and serves as a genetic blueprint is _______________, which is found in the _______________ of the cell.

3. The cytoplasm consists of _______________, which are specialized, membrane-bound intracellular compartments, and a gel-like mass known as _______________, which contains an elaborate protein network called the _______________.

Organelles (text page 18)

Contents

Section Synopsis

Five types of organelles are found in most cells: endoplasmic reticulum, Golgi complex, lysosomes, peroxisomes, and mitochondria. The endoplasmic reticulum is a single complex membranous network consisting of numerous interconnected tubules and flattened sacs that enclose a single fluid-filled space, or lumen. The primary function of the endoplasmic reticulum (ER) is to serve as a factory for synthesizing proteins and lipids to be used for: (1) the production of new cellular components, particularly cell membranes; and (2) secretion of special products such as enzymes and hormones to the exterior of the cell. There are two types of endoplasmic reticulum - rough endoplasmic reticulum, which is studded with ribosomes, and smooth endoplasmic reticulum, which lacks ribosomes. The rough endoplasmic reticular ribosomes synthesize proteins, which are threaded into the ER lumen so that they are separated from the cytosol. Also entering the lumen are lipids produced within the membranous walls of the ER. Synthesized products move from the rough ER to the smooth ER, where they are packaged and discharged to begin on their journey to their final destination. Transport vesicles are formed as a portion of the smooth ER "buds off", containing a collection of newly synthesized proteins and lipids wrapped in smooth ER membrane. In some specialized cells the smooth endoplasmic reticulum has additional responsibilities, such as storage of calcium.

The Golgi complex, which consists of stacks of flattened, membrane-bound sacs, serves a two-fold function: (1) it acts as a refining plant for modifying into a finished product the newly synthesized molecules delivered to it in crude form from the endoplasmic reticular factory; and (2) it sorts, packages, and directs molecular traffic to appropriate intracellular and extracellular destinations. Transport vesicles from the ER fuse with and dump their contents into the innermost sac of a Golgi stack. As the newly synthesized raw materials move through the layers of a Golgi stack by unknown means, specific biochemical modifications are accomplished to produce the final products. These products are packaged and discharged by the Golgi complex in different ways, depending on whether they are destined for various intracellular sites or for export from the cell.

Each cell contains several hundred lysosomes, which are membrane-bound sacs that contain powerful hydrolytic (digestive) enzymes. Serving as the intracellular digestive system, lysosomes perform the following functions: (1) destruction of phagocytized foreign material such as bacteria; (2) demolition of worn-out cell parts to make way for new replacement parts; (3) elimination of the entire cell if it is severely damaged or dead; (4) programmed cell death during embryonic development; and (5) regression of tissue when it is no longer needed.

Peroxisomes, small membrane-bound sacs containing powerful oxidative enzymes and catalase, are specialized for carrying out particular oxidative reactions, including certain detoxification activities. They generate hydrogen peroxide in the process, which they both use for further oxidative activity as well as destroy by means of catalase before the hydrogen peroxide can accumulate to dangerous levels.

The rod- or oval-shaped mitochondria are the energy organelles of the cell. They house the enzymes of the citric-acid cycle and electron-transport chain, which efficiently convert the energy in food molecules to the usable energy stored within ATP molecules. During this process, which is known as oxidative phosphorylation, the mitochondria utilize molecular oxygen and produce carbon dioxide and water as by-products. When O_2 is present (i.e., under aerobic conditions), oxidative phosphorylation generates a total net yield of thirty-six ATPs per molecule of glucose processed. In contrast, when O_2 is not available (i.e., under anaerobic conditions), only two ATPs are produced per molecule of glucose processed by glycolysis within the cytosol. The body's cells use ATP as an energy source for synthesis of new chemical compounds, for membrane transport, and for mechanical work.

Learning Check (Answers on p. A-3)

A. True/False

T/F 1. The endoplasmic reticulum is one continuous organelle consisting of many tubules and cisternae.

T/F 2. Coated vesicles enclose a representative mixture of proteins present in the Golgi sac before budding off.

T/F 3. Sometimes healthy organelles and/or cells are attacked by lysosomal enzymes.

T/F 4. Inadvertent rupture of lysosomal membranes inevitably leads to complete self-destruction of the cell.

T/F 5. Mitochondria are presumably descendants of primitive bacterial cells.

B. Fill-in-the-blank

1. The ribosomes of the rough ER synthesize _______________, whereas its membranous walls contain enzymes essential for the synthesis of _______________.

2. The signal-recognition protein recognizes both the _______________ on the ribosome and the _______________ on the ER to deliver the proper ribosome to the proper site on the rough ER for binding.

3. The _______________ ER is the central packaging and discharge site for molecules to be transported from the ER.

4. Transport vesicles from the _______________ fuse with and enter the _______________ for modification and sorting.

5. Products destined for intracellular transport are packaged in_______________, whereas products for export are packaged in_______________.

6. The process in which a secretory vesicle fuses with the plasma membrane, then opens up and extrudes its contents to the exterior is known as ______________.

7. Lysosomes contain (what type of) ______________ enzymes.

8. Foreign material to be attacked by lysosomal enzymes is brought into the cell by the process of ______________.

9. The (what kind of) _____________ enzymes within the peroxisomes primarily detoxify various wastes produced within the cell, generating __________________ in the process.

10. ______________, an enzyme found in peroxisomes, decomposes potentially toxic hydrogen peroxide.

11. The universal energy carrier of the body is ______________.

C. Matching

____ 1. synthesizes proteins used to construct new cell membrane

____ 2. synthesizes proteins used intracellularly within the cytosol

____ 3. synthesizes secretory proteins such as enzymes or hormones

a. free ribosome

b. rough ER-bound ribosome

D. Matching

____ 1. takes place in the mitochondrial matrix

____ 2. produces H_2O as a by-product

____ 3. rich yield of ATP

____ 4. takes place in the cytosol

____ 5. processes acetyl COA

____ 6. located in the inner mitochondrial membrane lining the cristae

____ 7. converts glucose into two pyruvic acid molecules

____ 8. utilizes molecular oxygen

a. glycolysis

b. citric-acid cycle

c. electron-transport chain

Cytosol and Cytoskeleton (text page 33)

Contents

The cytosol is important in intermediary metabolism, ribosomal protein synthesis, and storage of fat and glycogen. p. 33

Microtubules are essential for maintaining asymmetrical cell shapes and are important in complex cell movements. p. 34

Microfilaments are important in cellular contractile systems and as mechanical stiffeners. p. 38

Intermediate filaments are important in regions of the cell subject to mechanical stress. p. 40

The microtrabecular lattice provides a dynamic structural framework. p. 40

Section Synopsis

The cytosol contains the enzymes involved in intermediary metabolism and the ribosomal machinery essential for synthesis of these enzymes as well as other cytosolic proteins. Furthermore, many cells store within the cytosol unused nutrients in the form of glycogen granules or fat droplets. Pervading the cytosol is the cytoskeleton, which serves as the "bone and muscle" of the cell. The four types of cytoskeletal elements - microtubules, microfilaments, intermediate filaments, and microtrabecular lattice - are each composed of different proteins and perform different roles. Collectively, the cytoskeletal elements give the cell shape and support, enable it to arrange and move its internal structures as needed, and, in some cells, allow movement between the cell and its environment.

Learning Check (Answers on p. A-4)

A. Multiple choice

Which of the following is <u>not</u> characteristic of the cytoskeleton?

1. The cytoskeleton supports the plasma membrane and is responsible for the particular shape, rigidity, and spatial geometry of each different cell type.
2. The cytoskeleton probably plays a role in regulating cell growth and division.
3. The cytoskeletal elements are all rigid, permanent structures.
4. The cytoskeleton is responsible for cell contraction and cell movements.
5. The cytoskeleton supports and organizes the ribosomes, mitochondria, and lysosomes.

B. Matching

____ 1. hairlike motile protrusions

____ 2. found in hair cells of inner ear

____ 3. sweep mucus and debris out of respiratory airways

____ 4. stiffen the brush border of intestine and kidney cells

____ 5. enable sperm to move

____ 6. whip-like appendage

____ 7. guide egg to oviduct

a. flagella

b. cilia

c. actin bundles

C. Matching

____ 1. largest of cytoskeletal elements

____ 2. present in parts of the cell subject to mechanical stress

____ 3. smallest element(s) visible with a conventional electron microscope

____ 4. consist(s) of actin

____ 5. organize(s) the glycolytic enzymes in a sequential alignment

____ 6. form(s) the mitotic spindle

____ 7. essential for creating and maintaining an asymmetrical cell shape

____ 8. composed of tubulin

____ 9. provide(s) a pathway for axonal transport

____ 10. visible only with high voltage electron microscope

____ 11. play(s) a key role in muscle contraction

____ 12. slide past each other to cause ciliary bending

a. microtubules

b. microfilaments

c. intermediate filaments

d. microtrabecular lattice

D. True/False

T/F 1. Centrioles and basal bodies are identical in structure and may be interconvertible.

T/F 2. Amoeboid movement is accomplished by transitions of the cytosol between a gel and a sol state as a result of alternate assembly and disassembly respectively of actin filaments.

T/F 3. The protective, waterproof outer layer of skin is formed by the tough skeleton of the microtrabecular lattice that persists after the surface skin cells die.

Nucleus (text page 42)

Contents

Section Synopsis

The nucleus contains deoxyribonucleic acid (DNA), which is the cell's genetic material. A DNA molecule is composed of two long, helically intertwined strands of nucleotides. Each nucleotide consists of a deoxyribose sugar molecule, a phosphate group, and a nitrogenous base. The sugar and phosphate are identical for all nucleotides, but the base differs. Each nucleotide base is able to form a weak bond with another complementary base specific for it. The two strands within a DNA molecule are held together by the bonds that form between their precisely ordered complementary bases.

DNA performs two essential functions: (1) it serves as a blueprint for assembling proteins, which in turn determine the structural and functional characteristics of the body's cells; and (2) by replicating itself during cell division, it perpetuates the genetic blueprint within all new body cells and passes on genetic information to future generations.

Three types of ribonucleic acid (RNA) participate in protein synthesis. DNA's genetic code is transcribed by messenger RNA. Each sequence of three bases represents instructions for directing synthesis of a particular amino acid. A segment of DNA that codes for the sequence of amino acids in a given protein is known as a gene. During gene transcription, the DNA strands temporarily separate and the gene segment being transcribed serves as a template for the assembly of a messenger RNA molecule. Because messenger RNA is formed by complementary base pairing with DNA, it carries within its base sequence instructions for assembling a specified protein according to the genetic code. Once formed, messenger RNA detaches from DNA and moves into the cytosol, where it attaches to a ribosome. Ribosomes contain ribosomal RNA, which "reads" the base-sequence code of messenger RNA and translates it into the appropriate amino-acid sequence during synthesis of the specified protein. Transfer RNA delivers the designated amino acids to the growing protein chain under construction by the ribosome. Gene transcription and protein translation are governed by hormones and other factors that are responsible for controlling the rate of synthesis of each of the body's various proteins.

When a cell is not dividing, its DNA molecules are wound around bead-shaped protein molecules and folded to form a partially condensed, thread-like structure known as chromatin. During cell division, DNA becomes further folded and supercoiled into the condensed form of chromosomes, which are readily visible under a light microscope. All body cells, except the germ cells (i.e., sperm and eggs), contain forty-six chromosomes, each of which consists of a single highly folded, condensed DNA molecule. A cell's forty-six chromosomes are arranged into twenty-three pairs.

There are two types of cell division - mitosis, which is carried out by most body cells, and meiosis, which is accomplished only by germ cells. Before a cell divides, every chromosome is duplicated as the two strands of each DNA molecule separate and each strand replicates the missing strand by complementary base pairing so that two identical DNA molecules are formed. During mitosis, as the cell divides each of the two daughter cells receives an identical set of forty-six chromosomes (or DNA molecules), the same as the parent cell had. During meiosis, the cell divides twice so that each of the four resultant daughter cells receives a half set of twenty-three chromosomes, one member of each chromosome pair. Union of a sperm and egg during fertilization results in the formation of a new cell with a unique assortment of forty-six chromosomes, half from the father and half from the mother.

Learning Check (Answers on p. A-4)

A. Fill-in-the-blank

1. The nucleus of the cell contains ______________, the cell's genetic material.

2. The four DNA nitrogenous bases are __________, __________, __________, and __________.

3. A __________ is a stretch of DNA that codes for the synthesis of a particular polypeptide (protein).

4. Somatic cells contain (how many) _______ chromosomes, whereas germ cells contain ______ chromosomes.

5. The bead-like proteins around which a DNA molecule is wound are known as ______________.

6. The decondensed working form of DNA used as a template for messenger RNA assembly is known as ______________.

7. Each different amino acid is specified by a ______________ that consists of a specific sequence of three bases in the DNA nucleotide chain.

8. The three steps of protein synthesis are ______________, ______________, and ______________.

9. Nuclear division in somatic cells is accomplished by __________, whereas nuclear division in germ cells is accomplished by __________.

10. Condensed duplicate strands of DNA still joined together at the centromere are known as ______________.

11. The two components of cell division are nuclear division and ______________.

12. ____________ is the time interval between cell divisions.

13. ____________ involves a physical exchange of chromosome material between nonsister chromatids within a tetrad.

14. The largest group of known gene-signaling factors in humans is ____________.

B. True/False

T/F 1. Each ribosome can synthesize only one specific type of protein.

T/F 2. Only one ribosome can translate the mRNA message at any given time.

T/F 3. Mutations are always deleterious.

C. Matching

____ 1. chromatin condenses and chromosomes become visible; mitotic spindle appears

____ 2. sister chromatids separate and move toward opposite poles of the spindle

____ 3. chromosomes align themselves at the cell's midline

____ 4. cytoplasm divides

a. metaphase

b. prophase

c. anaphase

d. telophase

D. Matching

____ 1. carries the coded message from nuclear DNA to a cytoplasmic ribosome

____ 2. carries amino acids to their designated site in a protein under construction

____ 3. essential component of the workbenches for protein synthesis

____ 4. involved in transcription of RNA

____ 5. reads the base-sequence of mRNA and translates it into the appropriate amino-acid sequence during protein synthesis

____ 6. contains codons

____ 7. contains anticodons

a. ribosomal RNA

b. messenger RNA

c. transfer RNA

E. Multiple choice

1. Which of the following statements concerning complementary base pairing is incorrect?

   a. Adenine pairs with cytosine; guanine pairs with thymine.

   b. Complementay pairing between exposed DNA bases and free RNA bases takes place during transcription.

   c. Replication of DNA occurs during interphase.

   d. During DNA replication, the two DNA strands separate and each strand directs the assembly of free nucleotides to form the missing strand as a result of the imposed pattern of complementary base pairing.

2. Which of the following statements concerning meiosis is incorrect?

   a. During meiosis I, the members of each homologous pair of chromosomes line up to form a tetrad.

   b. Meiosis I yields two daughter cells, each containing twenty-three chromosomes consisting of two sister chromatids.

   c. During meiotic division, specialized germ cells do not undergo chromosome replication before the nucleus divides so that two daughter cells are formed, each containing a half set of twenty-three chromosomes.

   d. During meiosis II, sister chromatids separate for the first time.

Chapter in Perspective **(text page 57)**

C H A P T E R 3

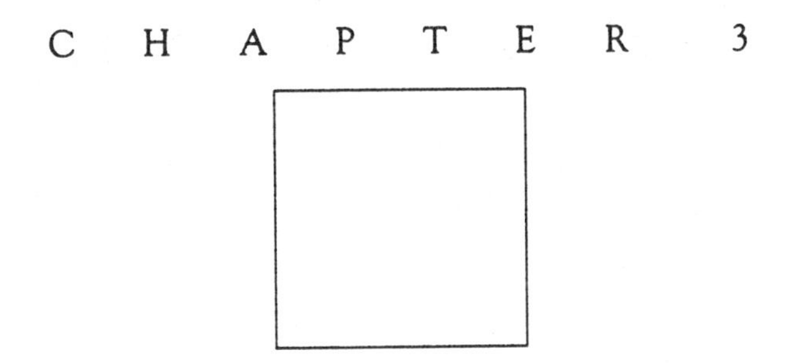

PLASMA MEMBRANE AND MEMBRANE POTENTIAL

Introduction (text page 58)

Membrane Structure and Composition (text page 59)

Contents

Section Synopsis

All cells are bounded by a plasma membrane, a thin lipid bilayer in which proteins are interspersed and to which carbohydrates are attached on the outer surface. The electron microscopic appearance of the plasma membrane as a trilaminar structure (two dark lines separated by a light interspace) is believed to be caused by the arrangement of the molecules composing it. The phospholipids orient themselves to form a bilayer with a hydrophobic interior (light interspace) sandwiched between the hydrophilic outer and inner surfaces (dark lines). This lipid bilayer forms the structural boundary of the cell, serving as a barrier for water soluble substances and being responsible for the fluid nature of the membrane. Cholesterol molecules tucked between the phospholipids contribute to the fluidity and stability of the membrane.

Proteins of various types either penetrate the thickness of the lipid bilayer or are attached to one of the membrane surfaces. These proteins, the type and distribution of which varies among cells, serve as: channels for passage of small ions across the membrane; carriers for transport of specific substances in or out of the cell; receptor sites for detecting and responding to chemical messengers that alter cell function; membrane-bound enzymes that govern specific chemical reactions; and a support meshwork on the inner surface to help maintain cell shape in association with the cytoskeleton.

The role of membrane carbohydrates is largely unknown, but it is speculated that the sugar chains projecting from glycoproteins are important in recognition of "self" in sophisticated cell-to-cell interactions such as tissue formation and immune processes. The asymmetrical arrangement of the plasma membrane components confers a sidedness on the membrane so that different types of activity take place at the inner and external surfaces.

Learning Check (Answers on p. A-6)

A. True/False

T/F 1. The nonpolar tails of the phospholipid molecules bury themselves in the interior of the plasma membrane.

T/F 2. The hydrophobic regions of the molecules composing the plasma membrane correspond to the two dark layers of this structure visible under an electron microscope.

T/F 3. The inner and outer layers of the plasma membrane are identical in composition and function.

B. Matching

a = lipid bilayer   b = proteins   c = carbohydrates

____ 1. channel formation

____ 2. barrier to passage of water-soluble substances

____ 3. receptor sites

____ 4. membrane fluidity

____ 5. recognition of "self"

____ 6. membrane-bound enzymes

____ 7. structural boundary

____ 8. carriers

Membrane Receptors and Postreceptor Events (text page 62)

Contents

Binding of chemical messengers to membrane receptors brings about a wide range of responses in different cells through use of only a few remarkably similar pathways. p. 62

Section Synopsis

Attachment of the first messenger (an extracellular chemical messenger such as a hormone) to a membrane receptor initiates one of several related intracellular pathways to bring about the desired response. The two major methods by which chemical messengers trigger cellular responses are: (1) by opening or closing specific channels, or (2) by activating an intracellular messenger (the second messenger). Two commonly employed second messengers are cyclic AMP and Ca^{++}. Once activated, these second messengers both initiate a similar cascade of intracellular events that ultimately lead to a change in shape and function of particular proteins to cause the appropriate cellular response. Variation in response among cells occurs despite the often widespread distribution of a single chemical messenger and despite the similarity of intracellular pathways employed because: (1) different cell types are equipped with different sets of receptors that can bind with only selected types of messengers from among the many that might come into contact with each of the cells; and (2) various cell types contain different intracellular proteins, each of which responds uniquely to an identical second messenger.

Learning Check (Answers on p. A-6)

A. Fill-in-the-blank

1. The two general ways in which interaction of a chemical messenger with a membrane receptor can bring about the desired intracellular responses are_________________ and ____________________.

2. The two major second messengers are ______ and _______.

B. Matching

____	1. activates cAMP	a. calmodulin
____	2. brings about mobilization of Ca^{++} from intracellular stores	b. adenylate cyclase
		c. phospholipase C
____	3. activated by Ca^{++}	d. protein kinase
____	4. activated by cAMP	

C. True/False

T/F 1. Second messenger systems ultimately bring about the desired cellular response by inducing a change in the shape and function of particular intracellular proteins.

T/F 2. One extracellular messenger molecule can ultimately influence the activity of only one protein molecule within the cell.

Cell-to-Cell Adhesions (text page 67)

Contents

Section Synopsis

The cells of the body are specifically organized and held together within tissues, and tissues are further arranged into organs. Special cells locally secrete a complex extracellular matrix, which serves as a biological glue between the cells of a tissue. This matrix consists of three major types of fibrous proteins embedded in a watery gel. The watery gel provides a passageway for diffusion of water-soluble particles between cells and the blood. The fibrous proteins include collagen, which confers tensile strength; elastin, which provides elasticity; and fibronectin, which promotes adhesion.

Many cells are further joined by specialized cell junctions. Desmosomes, which are fibrous interconnections that act like spot rivets to mechanically hold cells together, are especially important in tissues subject to a great deal of stretching. Tight junctions actually fuse cells together to seal off passage between cells, thereby permitting only regulated passage of materials through the cells. These impermeable junctions are found in the epithelial sheets that separate compartments with very different chemical compositions. Cells joined by gap junctions are connected by small channels that permit exchange of ions and small molecules between the cells. This movement plays a key role in the spread of electrical activity to synchronize contraction in heart and smooth muscle.

Learning Check (Answers on p. A-6)

A. Matching

____	1. provides tensile strength	a. collagen
____	2. provides a pathway for diffusion of water soluble particles between cells	b. elastin
		c. fibronectin
____	3. promotes cell adhesion	d. watery gel
____	4. enables tissue to stretch and recoil.	

B. Multiple Choice

Indicate the type of cell junction described in the question using the following answer code:

a = gap junction   b = tight junction   c = desmosome

____ 1. adhering junction

____ 2. impermeable junction

____ 3. communicating junction

____ 4. consists of connexons, which permit passage of ions and small molecules between cells

____ 5. consists of interconnecting fibers, which spot rivet adjacent cells

____ 6. consists of an actual fusion of proteins on the outer surfaces of two interacting cells

____ 7. important in tissues subject to mechanical stretching

____ 8. important in synchronizing contractions within heart and smooth muscle by allowing spread of electrical activity between the cells composing the muscle mass

____ 9. important in preventing passage between cells in epithelial sheets that separate compartments of two different chemical compositions

Membrane Transport **(text page 70)**

Contents

Lipid-soluble substances and small ions can passively diffuse through the plasma membrane down their electrochemical gradient. p. 70

Osmosis is the net diffusion of water down its own concentration gradient. p. 72

Special mechanisms are used to transport selected molecules unable to cross the plasma membrane on their own. p. 76

A Closer Look at Exercise Physiology - Exercising Muscles Have a "Sweet Tooth" p. 79

Section Synopsis

Materials can pass between the ECF and ICF by the following pathways. Nonpolar (lipid-soluble) molecules of any size can dissolve in and pass through the lipid bilayer. Small ions traverse through protein channels specific for them. Movement of particles through these pathways occurs passively down electrochemical gradients. Osmosis is a special case of water moving down its own concentration gradient.

Other substances can be selectively transferred across the membrane by specific carrier proteins, being moved either down a concentration gradient without the need for energy expenditure (facilitated diffusion) or moved against a concentration gradient at the expense of cellular energy (active transport). Carriers can move a single substance in one direction, two substances in opposite directions, or two substances in the same direction. Primary active transport requires the direct utilization of ATP to drive the pump, whereas secondary active transport is driven by an ion concentration gradient established by a primary active transport system. Carrier mechanisms are important for transfer of small polar molecules and for selected movement of ions.

Large polar molecules and multimolecular particles can leave or enter the cell by being wrapped in a piece of membrane to form vesicles that can be internalized (endocytosis) or externalized (exocytosis). Most vesicular traffic, as well as carrier transport across the membrane, involves attachment of the substance to be transported to specific protein binding sites within the plasma membrane. Cells are differentially selective in what enters or leaves because of variability in the number and kind of channels, carriers, and binding sites they possess. Large polar molecules (too large for channels and not lipid soluble) for which there are no special transport mechanisms are unable to permeate.

Learning Check (Answers on p. A-6)

A. True/False

T/F 1. When equilibrium is achieved and no net diffusion is taking place, there is no movement of molecules.

T/F 2. Phosphorylation of a carrier can alter the affinity of its binding sites, accompanied by a change in its conformation.

B. Multiple Choice

Use the following answer code to indicate the direction of net movement in each case:

a = movement from high to low concentration

b = movement from low to high concentration

____ 1. simple passive diffusion

____ 2. facilitated diffusion

____ 3. primary active transport

____ 4. Na^+ during secondary active transport

____ 5. co-transported molecule during secondary active transport

____ 6. water with regards to the water concentration gradient during osmosis

____ 7. water with regards to the solute concentration gradient during osmosis

C. Fill-in-the-Blank

1. Engulfment of a small volume of fluid by a cell is known as __________, whereas cellular ingestion of a multimolecular particle is referred to as __________. Collectively these processes are termed __________.
2. A rise in the cytosolic concentration of __________ induces fusion of an exocytotic vesicle with the plasma membrane.

Membrane Potential (text page 83)

Contents

Membrane potential refers to a separation of opposite charges across the plasma membrane. p. 83

Membrane potential is primarily due to differences in distribution and membrane permeability of sodium, potassium, and large intracellular anions. p. 86

Section Synopsis

All cells have a membrane potential, which is a separation of opposite charges across the plasma membrane. The Na^+-K^+ pump makes a small direct contribution to membrane potential through its unequal transport of positive ions; it transports more Na^+ ions out than K^+ ions in. The primary role of the Na^+-K^+ pump, however, is to actively maintain a greater concentration of Na^+ outside the cell and a greater concentration of K^+ inside the cell. These concentration gradients tend to passively move K^+ out of the cell and Na^+ into the cell. Because the resting membrane is much more permeable to K^+ than to Na^+, substantially more K^+ leaves the cell than Na^+ enters. This results in an excess of positive charges outside the cell and leaves an unbalanced excess of negative charges inside in the form of large protein anions (A^-) that are trapped within the cell. No further net movement of K^+ and Na^+ takes place when the resting membrane potential of -70 mV is achieved, because any further leaking of these ions down their concentration gradients is quickly reversed by the Na^+-K^+ pump. The distribution of Cl^- across the membrane is passively driven by the established membrane potential so that Cl^- is concentrated in the ECF.

Learning Check (Answers on p. A-7)

A. Fill-in-the-Blank

1. __________ refers to a separation of opposite charges across the membrane.

2. At the equilibrium potential for an ion, its ______________ gradient is exactly counterbalanced by its electrical gradient.

3. At resting membrane potential, there is a slight excess of_________(positive/negative) charges on the inside of the membrane, with a corresponding slight excess of ___________charges on the outside.

B. True/False

T/F 1. A potential of +30 mV is larger than a potential of -70mV.

T/F 2. The Na^+-K^+ pump is directly responsible for separating sufficient charges through its unequal pumping to establish a resting membrane potential of -70 mV.

T/F 3. At resting membrane potential, passive and active forces exactly balance each other so there is no net movement of ions across the membrane.

C. Indicate the various roles of the following ions using the answer code below :

$a = Na^+$　　　$c = A^-$

$b = K^+$　　　$d = Cl^-$

____ 1. cation in greatest concentration in the ICF

____ 2. cation in greatest concentration in the ECF

____ 3. anion in greatest concentration in the ICF

____ 4. anion in greatest concentration in the ECF

____ 5. ion whose equilibrium potential is greater than the resting membrane potential

____ 6. ion whose equilibrium potential is opposite in charge of the resting membrane potential

____ 7. ion whose equilibrium potential is exactly equal to the resting membrane potential

____ 8. cation to which the membrane is most permeable under resting conditions

____ 9. anion to which the membrane is impermeable

___ 10. ion that has the predominant influence on the resting membrane potential

___ 11. ion that is actively transported out of the cell

___ 12. ion that is actively transported into the cell

___ 13. ion whose concentration gradient is established by the membrane potential

Chapter in Perspective **(text page 90)**

CHAPTER 4

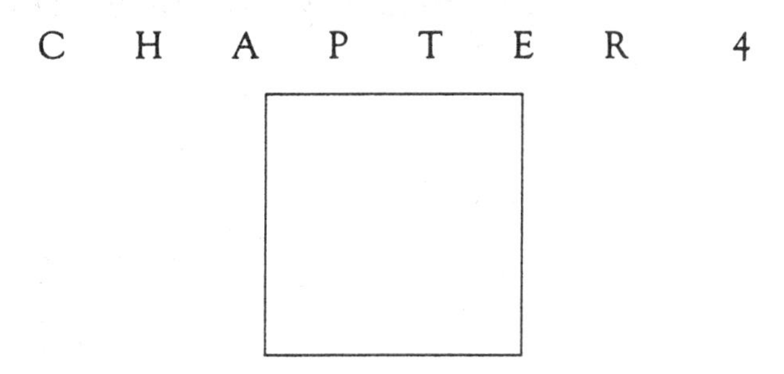

NEURONAL PHYSIOLOGY

Introduction (text page 92)

Electrical Signals: Graded Potentials and Action Potentials

(text page 93)

Contents

Fiber diameter also influences the velocity of action potential propagation. p. 101

Myelin plays a central role in several disorders. p. 102

The refractory period assures unidirectional propagation of the action potential and limits the frequency of action potentials. p. 103

Action potentials occur in all-or-none fashion. p. 104

Section Synopsis

Nerve and muscle cells are considered to be excitable tissues because they can rapidly alter their membrane permeabilities and thus undergo transient membrane potential changes when excited. The two kinds of potential change are: (1) graded potentials, which serve as short distance signals that quickly die off within a close range of the small patch of membrane where they are first triggered; and (2) action potentials, the long-distance signals.

During an action potential, depolarization of the membrane to threshold potential triggers sequential changes in permeability caused by conformational changes in voltage-gated channels. This brings about a brief reversal of membrane potential, with Na^+ influx being responsible for the rising phase (from -70 mV to +30 mV), followed by K^+ efflux during the falling phase (from peak back to resting potential). The Na^+ - K^+ pump gradually restores the ions that moved during the action potential to their original location to maintain the concentration gradients. Meanwhile, before an action potential returns to resting, it regenerates an identical new action potential right next to it via local current flow that brings the previously inactive area to threshold. This self-perpetuating cycle continues until the action potential has spread throughout the cell membrane in undiminished fashion. There are two types of action potential propagation: (1) conduction by local current flow in unmyelinated fibers, in which the action potential spreads along every portion of the membrane; and (2) the more rapid saltatory conduction in myelinated fibers, where the impulse jumps over the sections of the fiber covered with insulating myelin.

It is impossible to restimulate that portion of the membrane where the impulse has just passed until it has recovered from its refractory period. This assures the unidirectional propagation of action potentials away from the original site of activation.

Action potentials occur maximally in response to stimulation or not at all. Variable strengths of stimuli are coded by varying the frequency of action potentials, not their magnitude.

Learning Check (Answers on p. A-8)

A. Matching

_____ 1. behave in all or none fashion

_____ 2. magnitude of potential change varies with the magnitude of triggering response.

_____ 3. decremental spread away from original site

_____ 4. spreads throughout the membrane in nondiminishing fashion

_____ 5. serves as long distance signal

_____ 6. serves as short distance signal

a. graded potential

b. action potential

B. True/False

T/F 1. Conformational changes in channel proteins brought about by voltage changes are believed to be responsible for opening and closing Na^+ and K^+ gates during the generation of an action potential.

T/F 2. The Na^+ - K^+ pump restores the membrane to resting potential after it reaches the peak of an action potential.

T/F 3. Following an action potential, there is more K^+ outside the cell than inside because of the K^+ efflux.

C. Use the following answer code to answer the questions in this section:

a. increased P Na^+

b. decreased P Na^+

c. increased P K^+

d. decreased P K^+

e. Na^+ influx

f. Na^+ efflux

g. K^+ influx

h. K^+ efflux

1. Permeability change that occurs at threshold:______________

2. Two permeability changes that occur at the peak of an action potential: ____________ and _____________.

3. Ion movement responsible for the rising phase of the action potential: _____________.

4. Ion movement responsible for the falling phase of the action potential:_______________.

D. Fill-in the blank

1. List the two types of action potential propagation.

_____________________________ and _________________________.

Mark an * beside the fastest method of propagation.

2. The unidirectional propagation of action potentials away from the original site of activation is assured by the

______________________.

Synapses and Neuronal Integration (text page 104)

Contents

Section Synopsis

The primary means by which one neuron directly interacts with another neuron is through a synapse. An action potential in the presynaptic neuron triggers the release of a neurotransmitter, which combines with receptor sites on the postsynaptic neuron. This alters the latter cell in one of two ways. (1) The most typical response is the opening of voltage-gated channels. If both Na^+ and K^+ channels are opened, the resultant ionic fluxes cause an EPSP, a small depolarization that brings the postsynaptic cell closer to threshold. On the other hand, the likelihood that the postsynaptic neuron will reach threshold is diminished when an IPSP, a small hyperpolarization, is produced as a result of the opening of either K^+ or Cl^- channels, or both. (2) An alternate synaptic mechanism is activation of an intracellular second messenger system, such as cyclic AMP, by the neurotransmitter-receptor combination. Cyclic AMP can bring about channel opening or can have more prolonged effects within the cell, including alterations of the cell's genetic expression. Even though there are a number of different neurotransmitters, each synapse always releases the same transmitter to produce a given response when combined with a particular receptor. The response is terminated when the neurotransmitter is removed from the synaptic cleft by methods specific for the synapse.

The interconnecting synaptic pathways between various neurons are incredibly complex because of convergence of neuronal input and divergence of its output. A single neuron usually has many presynaptic inputs converging upon it that jointly control its level of excitability. This same neuron, in turn, diverges to synapse with and influence the excitability of many other cells. Each neuron thus has the task of computing an output to numerous other cells from a complex set of inputs to itself. Depending on the combination of signals it is receiving from its various presynaptic inputs, at any given time a neuron may react by: (1) firing action potentials along its axon; (2) remaining at rest and not passing any signals along; or (3) having its level of excitability reduced.

If the dominant activity is in its excitatory inputs, the postsynaptic cell is likely to be brought to threshold and have an action potential. This can be accomplished by either temporal summation (EPSPs from a repetitively firing presynaptic input occurring so close together in time that they add together) or spatial summation (adding of EPSPs occurring simultaneously from several different presynaptic inputs). Because the axon hillock has the lowest threshold, action potentials are initiated there. The frequency of action potentials reflects the magnitude of EPSP summation. If inhibitory inputs dominate, the postsynaptic potential will be brought farther than usual away from threshold. If excitatory and inhibitory activity to the postsynaptic neuron is balanced, the membrane will remain close to resting.

Numerous factors may alter synaptic effectiveness, some of which are built-in mechanisms to fine-tune neural responsiveness, some deliberate pharmacological manipulations to achieve a desired result, and some accidents caused by poisons or disease processes.

Learning Check (Answers on p. A-8)

A. Fill-in-the blank

1. A junction in which electrical activity in one neuron influences the electrical activity in another neuron by means of a neurotransmitter is called a ________________.
2. An action potential in a presynaptic neuron induces opening of voltage-gated ________________ channels in the synaptic knob, which triggers exocytosis of synaptic vesicles.
3. Summing of EPSPs occurring very close together in time as a result of repetitive firing of a single presynaptic input is known as________________.
4. Summing of EPSPs occurring simultaneously from several different presynaptic inputs is known as_________.
5. The ________________ is the site of action potential initiation in most neurons because it has the lowest threshold.
6. The neuronal relationship where synapses from many presynaptic inputs act upon a single postsynaptic cell is called ____________, whereas the relationship in which a single presynaptic neuron synapses with and thereby influences the activity of many postsynaptic cells is known as ______________.

B. Matching

_____	1. brings about increased permeability of the subsynatic membrane to both Na^+ and K^+	a. activation of an excitatory synapse
_____	2. causes a small hyperpolarization of the postsynaptic cell	b. activation of an inhibitory synapse
_____	3. causes a small depolarization of the postsynaptic neuron	
_____	4. brings about increased permeability of the subsynaptic membrane to K^+ or Cl^-	

C. True /False

T/F 1. The only way in which a neurotransmitter-receptor combination can influence the postsynaptic cell is to directly alter its permeability to specific ions.

T/F 2. Synapses permit two-way transmission of signals between two neurons.

T/F 3. A given synapse may produce EPSPs at one time and IPSPs at another time.

T/F 4. A balance of IPSPs and EPSPS will negate each other so that the grand postsynaptic potential is essentially unaltered.

Chapter in Perspective **(text page 113)**

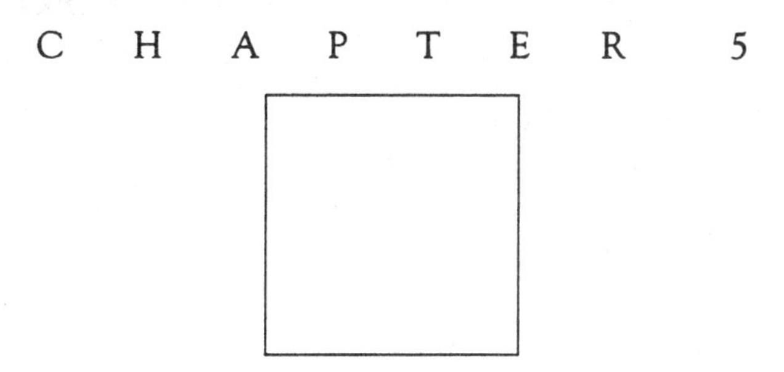

CHAPTER 5

Central Nervous System

Introduction (text page 116)

Our understanding of the brain is rudimentary because the brain is so complex and because there is no good model for study of its most sophisticated functions. p. 117

The brain is modified in response to environmental influences. p. 117

The nervous and endocrine systems have different regulatory responsibilities but share much in common. p. 118

The nervous system is organized into the central nervous system and the peripheral nervous system. p. 118

There are three classes of neurons. p. 119

Section Synopsis

The brain is the most complicated organ in the body and the least well understood. Our rudimentary knowledge of the brain is due not only to its complexity but also to limited opportunities for experimentation because of the uniqueness of many aspects of the human brain.

The nervous system is one of the two control systems of the body, the other being the endocrine system, with which the nervous system has much in common. Both systems utilize chemical messengers, sometimes even the same ones, to transmit signals to other cells. Furthermore, both systems have marked influences on each other. In general, the nervous system coordinates rapid responses, whereas the endocrine system is responsible for regulating activities that require duration rather than speed.

The nervous system is organized into the central nervous system (CNS), which includes the brain and spinal cord, plus the peripheral nervous system, which encompasses the nerve fibers carrying information to (afferent division) and from (efferent division) the CNS. Interacting networks of three classes of neurons - afferent neurons, efferent neurons, and interneurons - compose the excitable cells of the nervous sytem. Afferent neurons apprise the CNS of conditions in both the external and internal environment. Efferent neurons carry instructions from the CNS to effector organs, namely muscles and glands. Interneurons are responsible for integrating afferent information and formulating an efferent response, as well as for all higher mental functions associated with the "mind".

Learning Check (Answers on p. A-10)

A. Fill-in-the-blank

The _____________ system coordinates rapid responses of the body, whereas the _____________ system is responsible for regulating metabolic functions and activities that require duration rather than speed.

B. Matching

____ 1. consists of nerves carrying information between the periphery and CNS

____ 2. consists of brain and spinal cord

____ 3. division of the peripheral nervous system that transmits signals to the CNS

____ 4. division of the peripheral nervous system that transmits signals from the CNS

____ 5. supplies skeletal muscles

____ 6. supplies smooth muscle, cardiac muscle, and glands

a. somatic nervous system

b. autonomic nervous system

c. central nervous system

d. peripheral nervous system

e. efferent division

f. afferent division

C. Indicate which characteristics apply to each class of neuron by circling the appropriate letters using the answer code below (one characteristic may apply to more than one class of neurons).

a = characteristic of afferent neurons

b = characteristic of efferent neurons

c = characteristic of interneurons

1. have receptor at peripheral endings a b c

2. lie entirely within CNS a b c

3. lie primarily within PNS a b c

4. innervate muscles and glands a b c

5. cell body is devoid of of presynaptic inputs a b c

6. predominant type of neuron a b c

7. responsible for thoughts, emotions, memory, etc. a b c

Protection and Nourishment of the Brain (text page 120)

Contents

Neuroglia physically support the interneurons and help sustain them metabolically. p. 120

The delicate central nervous tissue is well-protected. p. 121

Three meningeal "mothers" provide protection and nourishment for the central nervous system. p. 121

The central nervous system is suspended in its own special cerebrospinal fluid. p. 121

A highly selective blood-brain barrier carefully regulates exchanges between the blood and brain. p. 123

The brain depends on constant delivery of oxygen and glucose by the blood. p. 125

Brain damage may occur in spite of protective mechanisms. p. 125

Headaches are seldom due to brain damamge. p. 125

Section Synopsis

Numerous glial cells within the CNS physically and metabolically support the neurons. Glial cells include astrocytes, oligodendrocytes, ependymal cells, and microglia, each of which has different specialized functions. The vital CNS tissue is provided with several protective devices. This is important because neurons are unable to undergo cell division to replace damaged cells. The delicate brain and spinal cord are wrapped in three layers of protective, nutritive membranes - the meninges (dura mater, arachnoid mater, and pia mater) and are further surrounded by a hard bony covering. A special cerebrospinal fluid flows within and around the CNS structures to cushion and provide them with an optimal environment for neuronal function. Protection against chemical injury is conferred by a highly selective blood brain barrier that limits access of blood-borne substances to the brain.

The brain depends on a constant blood supply for delivery of O_2 and glucose because it is unable to generate ATP in the absence of either of these substances. If brain damage occurs as a consequence of deprivation of blood supply, trauma, or a brain disorder, the resulting neurological deficit depends on the region of the brain involved, the extent of permanent damage, and the plasticity of the brain. Headaches, the most common form of pain, are only rarely associated with brain damage.

Learning Check (Answers on p. A-10)

A. Matching

_____ 1. brain phagocytes

_____ 2. line the brain ventricles

_____ 3. form the insulative myelin sheaths around axons in the CNS

_____ 4. main brain "glue"

a. astrocytes

b. oligodendrocytes

c. ependymal cells

d. microglia

B. Matching

_____ 1. tough, inelastic outer meningeal layer

_____ 2. venous blood draining from the brain empties here

_____ 3. tissue across which CSF is reabsorbed into the blood

_____ 4. site of formation of CSF

_____ 5. location of CSF as it surrounds the brain

a. arachnoid villi

b. subarachnoid space

c. choroid plexuses

d. dural sinuses

e. dura mater

C. True/False

T/F 1. The major function of CSF is to nourish the brain.

T/F 2. The brain can perform anaerobic metabolism in emergencies when oxygen supplies are low.

T/F 3. Cells forming the brain capillaries are joined by tight junctions that completely seal the capillary wall.

Cerebral Cortex (text page 126)

Contents

Newer, more sophisticated regions of the brain are piled on top of older, more primitive regions. p. 126

The cerebral cortex is an outer shell of gray matter covering an inner core of white matter. p. 128

The cerebral cortex is organized into layers and functional columns. p. 128

The four pairs of lobes in the cerebral cortex are specialized for different activities. p. 128

Other regions of the nervous system besides the primary motor cortex are important in motor control. p. 131

Somatotopic maps vary slightly between individuals and are dynamic, not static. p. 132

Language ability has several discrete components controlled by different regions of the cortex. p. 133

The association areas of the cerebral cortex are involved in many higher functions. p. 134

The cerebral hemispheres have some degree of specialization. p. 134

An electroencephalogram is a record of postsynaptic activity in cortical neurons. p. 134

Section Synopsis

The cerebral cortex is the outer shell of gray matter that caps an underlying core of white matter consisting of bundles of nerve fibers that interconnect various cortical regions with other areas. The cortex itself consists of neuronal cell bodies and their locally associated fibers organized into six discrete layers. These layers, which vary in thickness in different regions of the cortex, are arranged into functional vertical columns that serve as modular processing units responsible for specific activities.

Ultimate responsibility for many discrete functions is known to be localized in particular regions of the cortex, although each function is dependent on complex input-output interplay with other regions. The anatomical landmarks for cortical mapping are the four cortical lobes. Their basic functions are as follows: (1) The occipital lobe houses the visual cortex; (2) the auditory cortex is found in the temporal lobe; (3) the parietal lobe is primarily responsible for reception and perceptual processing of somatosensory input; and (4) voluntary motor movement is set into motion by frontal lobe activity. Language ability depends on the integrated activity of several discrete regions usually located only in the left hemisphere. The two primary language areas are Broca's area in the frontal lobe and Wernicke's area in the parietotemporal region. Wernicke's area is concerned with comprehension of spoken and written language and formulating the patterns of speech, which it relays to Broca's area. The latter is ultimately responsible for commanding the muscles necessary for speaking ability.

The areas of the cortex not specifically assigned to processing sensory input or commanding motor output or language ability are referred to as association areas. These areas provide an integrative link between diverse sensory information and purposeful action; they also play a key role in higher brain functions such as memory, planning, decision making, motivation, emotion, and so on. The exact cortical mapping for each individual is unique and dynamic, owing to genetic differences, brain plasticity, and use-dependent competition for cortical space.

There is some degree of hemispheric specialization. The left hemisphere is dominant for language as well as for logical, analytical, and sequential tasks. The right hemisphere excels in spatial, artistic, and musical abilities.

An electroencephalogram is a record of postsynaptic potential activity in the cortex. It is useful for diagnosing certain cerebral dysfunctions, distinguishing different stages of sleep, and declaring brain death in comatose patients.

Learning Check (Answers on p. A-10)

A. True/False

T/F 1. Gray matter refers to regions of the central nervous system composed primarily of densely packed cell bodies, whereas white matter consists of bundles of myelinated nerve fibers.

T/F 2. Damage to the left cerebral hemisphere brings about paralysis and loss of sensation on the left side of the body.

T/F 3. The hands and structures associated with the mouth have a disproportionately large share of representation in both the sensory and motor cortices.

T/F 4. The left cerebral hemisphere specializes in artistic and musical ability, whereas the right side excels in verbal and analytical skills.

T/F 5. The specific function a particular cortical region will carry out is permanently determined during embryonic development.

T/F 6. An electroencephalogram is a record of action potential activity in the cerebral cortex.

B. Matching

____	1. a thick band of axons transversing between the two hemispheres	a. temporal lobe
____	2. initial cortical processing for vision	b. Wernicke's area
____	3. initial cortical processing for hearing	c. somatosensory cortex
____	4. initial cortical processing for sensations arising from the surface of the body	d. limbic association cortex
____	5. programs complex sequences of movement	e. corpus callosum
____	6. triggers voluntary movement by activating motor neurons	f. primary motor cortex
____	7. responsible for speaking ability	g. occipital lobe
____	8. responsible for comprehension and formulation of coherent patterns of speech	h. Broca's area
____	9. primarily concerned with motivation and emotion	i. supplementary motor area
____	10. lesions in this area result in changes in personality and social behavior	j. prefrontal association cortex

Subcortical Structures and Their Relationship with the Cortex in Higher Brain Functions (text page 136)

The basal nuclei play an important inhibitory role in motor control. p. 136

The thalamus is a sensory relay station and is important in motor control. p. 136

The hypothalamus regulates many homeostatic functions. 137

The limbic system plays a key role in emotion and behavior.
p. 138

Motivated behavior and learning are believed to be influenced by rewards and punishments. p. 140

Norepinephrine, dopamine, and serotonin serve as neurotransmitters in motivated behavioral and emotional pathways.
p. 140

Learning is the acquisition of knowledge as a result of experiences. p. 141

Memory is laid down in stages. p. 141

Memory traces are present in multiple regions of the brain.
p. 142

Short-term and long-term memory appear to involve different molecular mechanisms. p. 143

Section Synopsis

The subcortical brain structures interact extensively with the cortex in the performance of their functions. The basal nuclei inhibit muscle tone; coordinate slow sustained postural contractions; and suppress useless patterns of movement. The thalamus serves as a relay station for preliminary processing of sensory input on its way to the cortex. The thalamus also reinforces voluntary motor activity initiated by the cortex. The hypothalamus regulates many homeostatic functions, in part through its extensive control of the autonomic nervous system and endocrine system. The limbic system, in close interaction with specific regions of the cortex, is responsible for basic inborn behavior patterns related to survival.

Learning further influences and modifies behavior. Intrinsic feelings of pleasure and punishment are incentives for learning. Individuals work to reinforce behaviors that lead to rewards and to avoid behaviors associated with punishment. Learning is the acquisition of knowledge as a result of experiences, whereas memory is its storage. The limbic system, cerebellum, temporal lobes and other cortical regions are strongly implicated in various aspects of formation, consolidation, and retention of memory traces. There are two types of memory: (1) a short-term memory with limited capacity and brief retention that is coded at least in part by temporary modifications in membrane channels; and (2) a long-term memory with large storage capacity and enduring memory traces, presumably involving relatively permanent structural or functional changes between already existing neurons.

Learning Check (Answers on p. A-11)

A. Fill-in-the blank

1. _______________ refers to the ability to direct behavior toward specific goals.
2. _______________ is the acquisition of knowledge as a consequence of experience.
3. The transfer and fixation of short-term memory traces into long-term memory stores is known as _______________.
4. The neural change responsible for retention or storage of knowledge is known as _______________.
5. The inability to recall recent past events following a traumatic event is known as _______________, whereas the inability to store new memories for later retrieval is called _____________.
6. _________________ is a decreased responsiveness to an indifferent stimulus that is repeatedly presented.
7. _________________ refers to increased reponsiveness to mild stimuli following a strong or noxious stimulus.

B. Matching

_____ 1. coordinates slow, sustained movements; inhibits muscle tone; suppresses unwanted patterns of movement

_____ 2. associated with transfer of new memories into long term storage; important in storage of verbal memories

_____ 3. relay station for synaptic input; reinforces voluntary movement initiated by the motor cortex

_____ 4. regulates the internal environment; links the autonomic nervous system and endocrine system

_____ 5. an interconnected ring of forebrain structures surrounding the brain stem; extensively involved with emotions, motivated behavior, and learning; contains "reward" and "punishment" centers

a. thalamus

b. hypothalamus

c. basal nuclei

d. limbic system

e. temporal lobes (and closely related structures)

C. Circle the appropriate letter using the following answer code to identify characteristics associated with memory types.

a = short-term memory   b = long-term memory

1. very large storage capacity   a b

2. limited-storage capacity   a b

3. site for initial deposition of new information   a b

4. takes longer to retrieve information from this store   a b

5. involves transient modifications in function of preexisting synapses   a b

6. probably involves relatively permanent functional or structural changes between existing neurons   a b

Cerebellum (text page 145) and Brain Stem (text page 147)

Contents

The cerebellum is important in balance as well as in planning and execution of voluntary movement. p. 145

The cerebellum and basal nuclei exert different influences on motor activity. p. 146

The brain stem is a vital link between the spinal cord and higher brain regions. p. 147

Sleep is an active process consisting of alternating periods of slow-wave and paradoxical sleep. p. 147

The sleep-wake cycle is probably controlled by interactions between three brain stem regions. p. 149

Section Synopsis

The cerebellum functions in the control of balance, eye movements, and muscle tone as well as in the coordination and planning of voluntary movement. It is especially important in smoothing out fast, phasic motor activities.

The brain stem is an important link between the spinal cord and higher brain levels. It is the origin of the cranial nerves; contains centers that control cardiovascular, respiratory, and digestive function; plays a role in motor control; modulates pain sensation; controls the overall degree of cortical alertness; and establishes the sleep-wake cycle. The prevailing state of consciousness is dependent on the cyclical interplay between an arousal system, a slow-wave sleep center, and a paradoxical sleep center all located in the brain stem.

Learning Check (Answers on p. A-11)

A. Fill-in-the blank

1. An ______________ tremor characterizes cerebellar disease.

2. _______________ refers to subjective awareness of surroundings and self.

B. Matching

______ 1. regulates muscle tone; compares intentions of higher centers with performance of muscles and corrects any errors, especially with rapid phasic movements

______ 2. plays a role in planning and initiation of voluntary activity; important for learning and remembering procedural motor tasks

______ 3. important for balance and eye movement

a. vestibulo-cerebellum

b. spino-cerebellum

c. cerebro-cerebellum

C. Indicate which characteristics apply to each type of sleep using the following answer code:

a = slow-wave sleep　　　b = paradoxical sleep

______ 1. rapid eye movements occur

______ 2. has four stages

______ 3. must go through this type of sleep first before entering the other type

______ 4. EEG pattern similar to a wide-awake, alert person

______ 5. dreaming occurs

______ 6. mental activity similar to waking-time thoughts

______ 7. inhibition of muscle tone

______ 8. frequent shifting of body position

______ 9. spend greatest percentage of time in this type of sleep

______ 10. heart and respiratory rate irregular

______ 11. hardest from which to arouse

______ 12. most apt to awaken from on own

Spinal Cord (text page 149)

Contents

The spinal cord extends through the vertebral canal and is connected to the spinal nerves. p. 149

The spinal cord is responsible for the integration of many basic reflexes. p. 152

A Closer Look at Exercise Physiology - Swan Dive or Belly Flop: It's a Matter of CNS Control p. 153

Section Synopsis

The spinal cord has two vital functions. First, it serves as the neuronal link between the brain and the peripheral nervous system. All communication up and down the spinal cord is located in well-defined, independent ascending and descending tracts in the cord's outer white matter. Second, it is the integrating center for spinal reflexes, including some of the basic protective and postural reflexes and those involved with emptying of the pelvic organs.

The components of a basic reflex arc include a receptor, an afferent pathway, an integrating center, an efferent pathway, and an effector. The centrally located gray matter of the spinal cord contains the interneurons interposed between the afferent input and efferent output as well as the cell bodies of efferent neurons. The afferent and efferent fibers, which carry signals to and from the spinal cord respectively, are bundled together into spinal nerves. These nerves are attached to the spinal cord in paired fashion throughout its length. They supply specific regions of the body. The afferent fibers within a particular spinal nerve are linked to receptors that respond to detectable changes within a given region; the efferent fibers in the nerve terminate on effectors (muscles or gland) within the same general region. Higher brain centers can exert considerable influence over spinal reflexes.

Learning Check (Answers on p. A-11)

A. Matching

_____ 1. location of ascending and descending tracts

_____ 2. location of cell bodies for efferent neurons

_____ 3. location of cell bodies for afferent neurons

_____ 4. location of short interneurons involved in integration of spinal reflexes

_____ 5. outer portion of spinal cord

______6. inner portion of spinal cord

a. gray matter

b. white matter

c. dorsal root ganglion

B. True/False

T/F 1. Information as to whether a finger was touching an ice cube or being hit by a hammer would be carried to the brain in different ascending tracts within the spinal cord.

T/F 2. A central bundle of interneuronal axons is known as a tract, whereas a peripheral bundle of afferent and efferent neuronal axons is called a nerve.

C. Fill-in-the-blank

1. Afferent fibers enter through the __________root of the spinal cord, whereas efferent fibers leave through the__________root.

2. List the five components of a basic reflex arc: 1.___________ 2.____________ 3.____________ 4.____________ 5.____________.

Chapter in Perspective (text page 156)

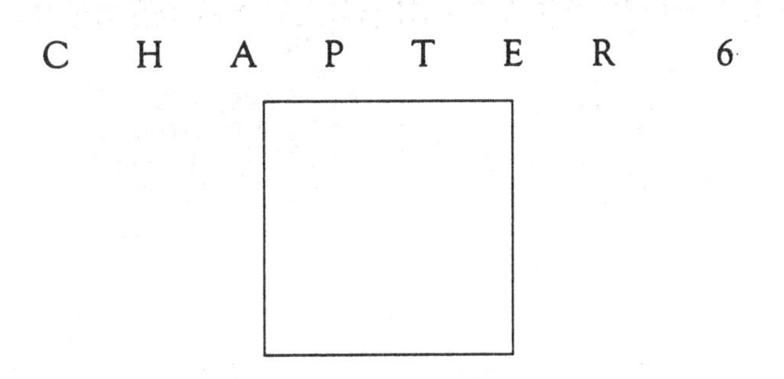

CHAPTER 6

PERIPHERAL NERVOUS SYSTEM: AFFERENT DIVISION; SPECIAL SENSES

Introduction (text page 158)

Receptors; Sensory Pathways (text page 159)

Contents

Section Synopsis

Perceptions, which represent the conscious interpretation of the external world based on processing of sensory input, do not replicate reality. Only stimuli for which receptors are present can be detected. Furthermore, central nervous system processing modifies the input, sifting out extraneous information and concentrating on important cues. Afferent input is necessary for arousal, perception, and determination of efferent output.

Receptors are specialized structures that respond to particular stimuli, translating the energy forms of the stimuli into electrical signals, the language of the nervous system. There are discrete labeled-line pathways from the receptors to the CNS so that information about the type and location of the stimuli can be deciphered by the CNS, even though all the information arrives in the form of action potentials. Information that reaches the conscious level is called sensory information, which includes somatic sensation (body awareness) and the special senses (seeing, hearing, tasting, and smelling).

Stimulation of a receptor produces a graded receptor potential, typically a depolarizing response caused by Na^+ influx. The strength and rate of change of the stimulus are reflected in the magnitude of the receptor potential, which in turn determines the frequency of action potentials generated in the afferent neuron. The magnitude of the receptor potential is also influenced by the extent of receptor adaptation, which refers to a reduction in receptor potential in spite of sustained stimulation. Tonic receptors adapt slowly or not at all, thus providing continuous information about the stimuli they monitor. Phasic receptors adapt rapidly and frequently exhibit off responses, thereby providing information about changes in the energy form they monitor.

The smaller the receptive field of sensory neurons within a given region, the greater the region's discriminative ability. Lateral inhibition within the CNS sharpens contrast and facilitates localization of a stimulus.

Learning Check (Answers on page A-12)

A. Fill-in-the-blank

1. Conversion of the energy forms of stimuli into electrical energy by the receptors is known as______________ .
2. The type of stimulus to which a particular receptor is most responsive is called its ___________________.

B. True/False

T/F 1. The sensation perceived depends on the type of receptor stimulated rather than on the type of stimulus.

T/F 2. Humans have receptors to detect all stimulus modalities in the environment.

T/F 3. All afferent information is sensory information.

C. Indicate how coding of the following types of sensory information is accomplished.

1. Type of stimulus________________________________

2. Location of stimulus_____________________________

3. Intensity of stimulus____________________________

D. Matching

_____1. generates action potentials in afferent neuron by opening voltage-gated Na^+ channels.

_____2. generates action potentials in afferent neuron by opening messenger-gated channels

_____3. provides continuous information about the stimuli being monitored

_____4. provides information about changes in the energy form being monitored

a. tonic receptors

b. phasic receptors

c. specialized receptor ending of afferent neuron

d. separate cell serving as a receptor

Pain (text page 165)

Contents

Stimulation of nociceptors elicits the perception of pain plus motivational and emotional responses. p. 165

The brain has a built-in analgesic system. p. 166

Section Synopsis

Painful experiences are elicited by noxious mechanical, thermal, or chemical stimuli and consist of two components: the perception of pain coupled with emotional and behavioral responses to it. Locally released prostaglandins increase the sensitivity of nociceptors. Transmission of pain signals takes place over two afferent pathways: a fast pathway consisting of large, myelinated A-delta fibers that produce sharp, prickling pain; and a slow pathway composed of small, unmyelinated C fibers that carry dull, aching, persistent pain signals. Afferent pain fibers terminate on ascending pathways within the dorsal horn of the spinal cord. Central nervous system structures involved with processing and responding to pain input include the somatosensory cortex, thalamus, reticular formation, hypothalamus, and limbic system. Descending pathways that involve the periaqueductal gray matter and reticular formation use endogenous opiates to suppress the release of Substance P, the neurotransmitter from the afferent pain fiber terminal. This blocks further transmission of the pain signal, thus serving as a built-in analgesic system.

Learning Check (Answers on p. A-12)

A. Indicate the properties associated with each type of nociceptor by circling the appropriate letters using the answer code below:

a = mechanical nociceptor

b = thermal nociceptor

c = polymodal nociceptor

1. respond(s) to cutting, crushing, pinching a b c
2. respond(s) to irritating chemicals a b c
3. respond(s) to temperature extremes a b c
4. transmits its signals via A-delta fibers a b c
5. transmits its signals via C fibers a b c
6. activates slow-pain pathway a b c
7. activates fast-pain pathway a b c

B. Fill-in-the-blank

1. ________________, a neurotransmitter unique to pain fibers, is released from afferent pain terminals.

2. ________________ receptors bind with endorphins, enkephalins, and morphine.

C. Indicate which of the following regions are part of ascending pain pathways or descending analgesic pathways or both.

_____1. periaqueductal gray area

_____2. thalamus

_____3. reticular formation

a. part of ascending pain pathway

b. part of descending analgesic pathway

c. participates in both pain and analgesic pathways

D. True/False

T/F 1. Prostaglandins suppress the sensitivity of nociceptors.

T/F 2. Descending analgesic pathways are believed to suppress transmission in the pain pathways as they enter the spinal cord by blocking the release of substance P.

Eye: Vision **(text page 166)**

Contents

Accommodation increases the strength of the lens for near vision. p. 172

Light must pass through several retinal layers before reaching the photoreceptors. p. 174

Phototransduction by retinal cells converts light stimuli into neural signals. p. 174

Rods provide indistinct gray vision at night whereas cones provide sharp color vision during the day. p. 177

The sensitivity of the eyes can vary markedly through dark and light adaptation. p. 179

Color vision is dependent on the ratios of stimulation of the three cone types. p. 179

Visual information is separated and modified within the visual pathway before it is integrated into a perceptual image of the visual field by the cortex. p. 180

The thalamus and visual cortices elaborate the visual message. p. 182

Visual input goes to other areas of the brain not involved in vision perception. p. 183

Protective mechanisms help prevent eye injuries. p. 183

Section Synopsis

The eye is a specialized structure housing the light-sensitive receptors essential for vision perception - namely, the rods and cones found in its retinal layer. The iris controls the size of the pupil, thereby adjusting the amount of light permitted to enter the eye. The cornea and lens are the primary refractive structures that bend the incoming light rays to focus the image on the retina. The cornea contributes most to the total refractive ability of the eye. The strength of the lens can be adjusted through action of the ciliary muscle to accommodate for differences in near and far vision.

Rods and cones are activated when the photopigments they contain differentially absorb various wavelengths of light, which causes the two photopigment components, the vitamin A derivative retinal and the protein opsin, to dissociate. This dissociation activates a second messenger system within the photoreceptor, which brings about closure of Na^+ channels and a subsequent hyperpolarization of the receptor synaptic terminal. The result is a disinhibition, or in effect, excitation, of the next cells in the retinal neural chain.

Differences in photopigment sensitivity and wiring patterns within the retina confer different properties on rod and cone visual pathways. Cones display high acuity but can be used only for day vision because of their low sensitivity to light. Different ratios of stimulation of three cone types by varying wavelengths of light leads to color vision. Rods provide only indistinct vision in shades of gray, but, because they are very sensitive to light, they can be used for night vision. The eyes are able to markedly adjust their sensitivity to light via dark and light adaptation, which depends on the relative amount of intact photopigment present in the rods and cones.

Several other retinal neuronal layers process and modify the visual message passed on from the rods and cones to increase image contrasts before the signals are sent to the brain. Various aspects of visual information are separated and processed in parallel pathways through hierarchical systems of cells, being sorted and sent to appropriate zones of the visual cortices by the thalamus. Cells within the primary and higher visual cortices range in complexity depending on the visual pattern to which they respond. It is not known how the various components of visual information are integrated into a perceptual whole image.

External eye muscles control eye movements to keep the image focused on the fovea, the central point of the retina with the capacity for the most distinct vision. The eyelids shutter the eye, tears wash it, and eyelashes trap air-borne particles to protect the eye from injury.

Learning Check (Answers on p. A-13)

A. True/False

T/F 1. Off-center ganglion cells increase their rate of firing when a beam of light strikes the periphery of their receptive field.

T/F 2. Short wavelengths of light are perceived in the red-orange color range.

T/F 3. Photoreceptors, bipolar cells, and ganglion cells all display action potentials.

T/F 4. During dark adaptation, rhodopsin is gradually regenerated to increase the sensitivity of the eyes.

T/F 5. The visual message detected by the retina is transmitted intact to the visual cortex.

T/F 6. An optic nerve carries information from the lateral and medial halves of the same eye, whereas an optic tract carries information from the lateral half of one eye and the medial half of the other.

T/F 7. Each half of the visual cortex receives information from the opposite half of the visual field as detected by both eyes.

T/F 8. Binocular vision enhances depth perception.

B. Matching

_____ 1. layer that contains the photoreceptors

_____ 2. point from which the optic nerve leaves the retina

_____ 3. forms the white part of the eye

_____ 4. thalamic structure that processes visual input

_____ 5. colored diaphragm of muscle that controls amount of light entering the eye

_____ 6. contributes most to refractive ability of the eye

_____ 7. supplies nutrients to the lens and cornea

_____ 8. produces aqueous humor

_____ 9. contains vascular supply for the retina and pigment that minimizes scattering of light within the eye

_____10. has adjustable refractive ability

_____11. portion of retina with greatest acuity

_____12. point at which fibers from the medial half of each retina cross to the opposite side

a. choroid

b. aqueous humor

c. fovea

d. lateral geniculate nucleus

e. cornea

f. retina

g. lens

h. optic disc; blind spot

i. iris

j. ciliary body

k. optic chiasm

l. sclera

C. Complete the statements by circling the correct response.

1. For near vision, the ciliary muscle (contracts or relaxes) so that the suspensory ligaments become (taut or slack). This allows the lens to (flatten or round up), which (increases or decreases) the strength of the lens.

2. When light of suitable wavelength strikes a photoreceptor, the photopigment (absorbs or reflects) the light, causing the photopigment to (dissociate or regenerate). This photopigment transformation acts through a second messenger system to cause Na^+ channels in the outer segment to (open or close). Subsequently the photoreceptor (depolarizes or hyperpolarizes), which results in a(n) (increase or decrease) in transmitter release from its synaptic terminal.

D. Circle the correct letter to indicate the properties of rods and cones using the following answer code:

a = rods   b = cones

1. used for day vision   a b
2. used for night vision   a b
3. confer color vision   a b
4. confer vision in shades of gray   a b
5. high acuity   a b
6. low acuity   a b
7. contain opsin and retinal   a b
8. much convergence in pathways   a b
9. little convergence in pathways   a b
10. three different types as a result of difference in photopigment content   a b

E. Matching

_____ 1.	eyeball too long	a. color blindess
_____ 2.	eyeball too short	b. night blindness
_____ 3.	corrected by cylindrical lens	c. glaucoma
_____ 4.	corrected by concave lens	d. hyperopia
_____ 5.	corrected by convex lens	e. diplopia
_____ 6.	corneal surface uneven	f. presbyopia
_____ 7.	images from two eyes not fused within cortex	g. myopia
_____ 8.	increased intraocular pressure	h. cataract
_____ 9.	opaque lens	i. astigmatism
_____10.	stiffened lens	
_____11.	Vitamin A deficiency	
_____12.	lack of a cone type	

Ear : Hearing and Equilibrium **(text page 183)**

Contents

Sound waves consist of alternate regions of compression and rarefaction of air molecules. p. 183

The external and middle ear convert air-borne sound waves into fluid vibrations in the inner ear. p. 186

Hair cells in the organ of Corti transduce fluid movements into neural signals. p. 188

Pitch discrimination depends on the place on the basilar membrane that vibrates; loudness discrimination depends on the amplitude of the vibration. p. 190

The auditory cortex is mapped according to tone. p. 190

Deafness is caused by defects either in conduction or neural processing of sound waves. p. 191

The vestibular apparatus detects position and motion of the head important for equilibrium and coordination of head and body movement. p. 192

Section Synopsis

The ear performs two unrelated functions: (1) hearing, which involves the external ear, middle ear, and cochlea of the inner ear; and (2) sense of equilibrium, which involves the vestibular apparatus of the inner ear. Hearing depends on the ear's ability to convert air-borne sound waves into mechanical deformations of receptive hair cells, thereby initiating neural signals. Sound waves consist of high-pressure regions of compression alternating with low-pressure regions of rarefaction of air molecules. The pitch (tone) of a sound is determined by the frequency of its waves and the loudness (intensity) by the amplitude of the waves. Sound waves are funneled through the external ear canal to the tympanic membrane, which vibrates in synchrony with the waves. Middle-ear bones bridging the gap between the tympanic membrane and the inner ear amplify the tympanic movements and transmit them to the oval window, whose movement sets up traveling waves in the cochlear fluid. These waves, which are at the same frequency as the original sound waves, set the basilar membrane in motion. Various regions of the membrane selectively vibrate more vigorously in response to different frequencies of sound. On top of the basilar membrane are the receptive hair cells of the organ of Corti, whose hairs are bent as the basilar membrane is deflected up and down in relationship to the overhanging stationary tectorial membrane in which the hairs are embedded. This mechanical deformation of specific hair cells in the region of maximal basilar membrane vibration is transduced into neural signals transmitted via the auditory nerve to the CNS for sound perception. The pitch of the perceived sound is determined by the place on the basilar membrane that vibrates maximally in response to the frequency of the incoming sound. Loudness discrimination depends on the amplitude of vibration of the basilar membrane. Sound localization is dependent on the fact that the ear closer to the sound source detects the sound sooner and more intensely than the other ear. For sounds directly in the front or to the rear, the shape of the pinna is useful in distinguishing the location.

The auditory pathway includes synaptic relays at the brain stem and thalamus, finally terminating in the auditory cortex of the temporal lobe. Each region of the auditory cortex is linked with a specific portion of the basilar membrane, so the cortex is mapped according to tone. Deafness occurs when the sound wave is not adequately transmitted to the fluid of the inner ear (conduction deafness), or when the fluid vibrations are not properly transduced into impulses perceived by the auditory cortex as sound (nerve deafness).

The vestibular apparatus in the inner ear consists of: (1) the semicircular canals, which detect rotational acceleration or deceleration in any direction; and (2) the utricle and saccule, which detect changes in the rate of linear movement in any direction and provide information important for determining head position in relationship to gravity. Neural signals are generated in response to mechanical deformation of hair cells caused by specific movement of fluid and related structures within these sense organs. This vestibular input regarding head motion and position is used by the vestibular nuclei in the brain stem to coordinate head movements for tracking fixed objects in the visual field as the head is turning and to initiate reflexes important in maintaining balance and posture.

Learning Check (Answers on p. A-13)

A. Indicate the characteristics associated with each part of the ear by circling the appropriate letters using the answer code below:

a = external ear

b = middle ear

c = cochlea in the inner ear

d = semicircular canals in the inner ear

e = utricle and saccule in the inner ear

1. air-filled a b c d e

2. fluid-filled a b c d e

3. contain(s) receptive hair cells a b c d e

4. concerned with hearing a b c d e

5. concerned with sense of equilibrium a b c d e

6. contain(s) the tympanic membrane, which vibrates in synchrony with sound waves that strike it a b c d e

7. contain(s) the ossicular system, which contributes to the amplification of the sound wave — a b c d e

8. contain(s) a cupula, which sways in the direction of endolymph movement, bending the embedded hair cells — a b c d e

9. provides information about the position of the head relative to gravity — a b c d e

10. detect(s) rotational acceleration or deceleration of the head — a b c d e

11. contain(s) otoliths in gelatinous mass, movement of which bends hair cells — a b c d e

12. contain(s) the organ of Corti whose hair cells are bent during vibration of the basilar membrane — a b c d e

13. is/are connected with the throat via the eustachian tube — a b c d e

14. provide(s) information useful for keeping the eyes focused on a fixed object even when the head is moving — a b c d e

15. part of the vestibular apparatus — a b c d e

16. consist(s) of three elongated spiral compartments — a b c d e

17. components aligned in each of three different planes all perpendicular to each other — a b c d e

B. Matching: characteristics of sound

_____ 1. determined by frequency of sound waves

_____ 2. dependent on overtones

_____ 3. determined by amplitude of sound waves

_____ 4. imparts different sounds to different instruments playing the same note

_____ 5. discrimination of this characteristic of a sound depends on the location of greatest basilar membrane vibration in response to the sound

_____ 6. discrimination of this characteristic of a sound depends on the amplitude of basilar membrane vibration

_____ 7. measured in Hertz

_____ 8. measured in decibels

a. pitch (tone)

b. intensity (loudness)

c. timbre (quality)

C. True/False

T/F 1. Sound waves consist of regions of high-pressure compression alternating with regions of low-pressure rarefaction of air molecules.

T/F 2. A 100 dB sound is 100 times louder than hearing threshold.

T/F 3. Mechanical deformation of the hairs of hair cells always results in depolarization of the hair cell synaptic terminal.

T/F 4. Displacement of the round window generates neural impulses that are perceived as sound sensations.

T/F 5. Hair cells in different regions of the organ of Corti and neurons in different regions of the auditory cortex are activated by different tones.

T/F 6. Exposure to very loud noises can result in partial conduction deafness.

T/F 7. Hearing aids are helpful in conduction deafness but not in nerve deafness.

T/F 8. The ossicular system transmits the vibrations of the tympanic membrane to the oval window, movement of which sets up traveling waves in the cochlear fluid.

T/F 9. The vestibular nuclei provide output important in maintaining balance and posture.

Chemical Senses: Taste and Smell **(text page 196)**

Contents

Taste sensation is coded by patterns of activity in various taste bud receptors. p. 196

Smell is the least understood of the special senses. p. 197

Section Synopsis

Taste and smell are chemical senses. In both cases, attachment of specific dissolved molecules to binding sites on the surface of the receptor membrane causes channel changes that lead to depolarizing receptor potentials, which, in turn, set up neural impulses that signal the presense of the chemical. Taste receptors are housed in taste buds on the tongue; olfactory receptors are located in a patch of mucosa in the upper part of the nasal cavity. Taste receptors are separate cells that synapse with terminal endings of afferent neurons; olfactory receptors are specialized endings of afferent neurons themselves. Taste receptors are turned over every ten days; olfactory receptors are generated anew every sixty days. Both sensory pathways include two routes: one to the limbic system for affective and behavioral processing, and one through the thalamus to the cortex for conscious perception and fine discrimination. Taste discrimination depends on the pattern of stimulation in all of the taste receptors, which vary in their sensitivity to each of the four primary tastes of sweet, sour, salty, and bitter. Odor discrimination is less well understood but presumably involves a similar mechanism.

Learning Check (Answers on p. A-14)

A. Matching: taste stimulants

_____1. salty	a. alkaloids
_____2. sour	b. acids
_____3. bitter	c. anything with chemical configuration similar to glucose
_____4. sweet	d. NaCl

B. Indicate which properties apply to taste and/or smell by circling the appropriate letter(s) using the following answer code:

a = applies to taste b = applies to smell

c = applies to both taste and smell

1. Receptors are separate cells that synapse with terminal endings of afferent neurons a b c

2. Receptors are specialized endings of afferent neunons a b c

3. Receptors are regularly replaced a b c

4. Specific chemicals in the environment attach to special binding sites on receptor surface leading to a depolarizing receptor potential a b c

5. Has two processing pathways: a limbic system route and a thalamic-cortical route a b c

C. True/False

T/F 1. Each taste receptor responds to just one of the four primary tastes.

T/F 2. The cortical gustatory area is located adjacent to the "tongue" region of the somatosensory cortex.

T/F 3. Normal breathing patterns directly bring odoriferous molecules in contact with the olfactory mucosa.

T/F 4. Molecules of similar smell have a similar chemical composition.

T/F 5. Rapid adaptation to odors results from adaptation of the olfactory receptors.

Chapter in Perspective **(text page 199)**

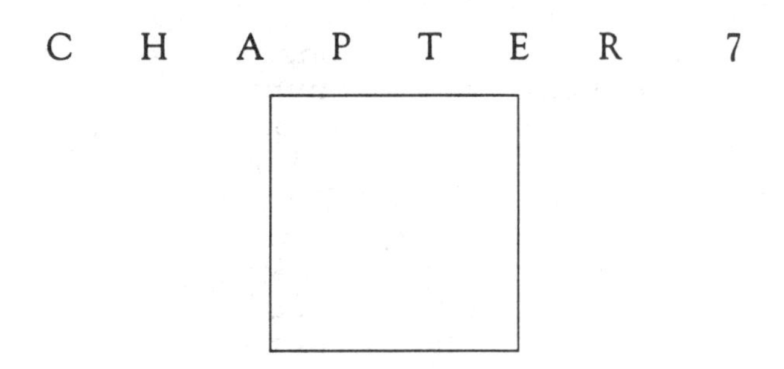

Peripheral Nervous System: Efferent Division

Introduction (text page 202)

Autonomic Nervous System (text page 203); and Somatic Nervous System (text page 208)

Contents

An autonomic nerve pathway consists of a two-neuron chain, with the terminal neurotransmitter differing between sympathetic and parasympathetic nerves. p. 203

The autonomic nervous system controls involuntary visceral-organ activities. p. 204

The sympathetic and parasympathetic nervous systems dually innervate most visceral organs. p. 205

There are several different types of membrane receptor proteins for each autonomic neurotransmitter. p. 207

Many regions of the central nervous system are involved in the control of autonomic activities. p. 208

Alpha motor neurons supply skeletal muscle. p. 208

Alpha motor neurons are the final common pathway. p. 208

Section Synopsis

The efferent division of the peripheral nervous system carries directives from the central nervous system to the effector organs. Cardiac muscle, smooth muscle, and exocrine glands are innervated by the autonomic nervous system, which is considered to be the involuntary branch of the peripheral efferent division. Skeletal muscle is innervated by the somatic nervous system, which is the voluntary branch of the efferent division.

The autonomic nervous system consists of two subdivisions - the sympathetic and parasympathetic nervous systems. An autonomic nerve pathway consists of a two-neuron chain. The preganglionic fiber originates in the CNS and synapses with the cell body of the postganglionic fiber in a ganglion outside of the CNS. The postganglionic fiber terminates on the effector organ.

All preganglionic fibers plus parasympathetic postganglionic fibers release acetylcholine. Sympathetic postganglionic fibers release norepinephrine. Variability in responsiveness of different tissues to the same neurotransmitter depends on specialization of the tissue cells, not on the properties of the messenger. Tissues innervated by the autonomic nervous system possess one or more of several different receptor types for the postganglionic chemical messengers.

A given autonomic fiber either excites or inhibits activity in the organ it innervates. Most visceral organs are innervated by both sympathetic and parasympathetic nerve fibers, which in general produce opposite effects in a particular organ. Dual innervation of visceral organs by both branches of the autonomic nervous system permits precise control over an organ's activity. The sympathetic system dominates in emergency or stressful situations and promotes responses that prepare the body for strenuous physical activity. The parasympathetic system dominates in quiet, relaxed situations and promotes body maintenance activities such as digestion.

Autonomic nervous system output is regulated by various regions of the CNS, including the spinal cord, medulla, hypothalamus, and frontal cortex.

The somatic nervous system consists of the axons of alpha motor neurons, which originate in the spinal cord and terminate on skeletal muscle. Acetylcholine, the neurotransmitter released from an alpha motor neuron, stimulates muscle contraction. Alpha motor neurons are the final common pathway by which various regions of the CNS, including the motor regions of the cortex, the basal nuclei, the cerebellum, the brain stem, and spinal cord, are able to exert control over skeletal muscle activity.

Learning Check (Answers on p. A-15)

A. Use the following answer code to identify the autonomic transmitter being described in each question.

a = acetylcholine b = norepinephrine

_____ 1. secreted by all preganglionic fibers

_____ 2. secreted by sympathetic postganglionic fibers

_____ 3. secreted by parasympathetic postganglionic fibers

_____ 4. secreted by the adrenal medulla

_____ 5. secreted by alpha motor neurons

_____ 6. binds to muscarinic or nicotinic receptors

_____ 7. binds to α or ß receptors

B. Indicate the characteristics of the two types of efferent output using the following answer code :

a = characteristic of the somatic nervous system

b = characteristic of the autonomic nervous system

_____ 1. composed of two-neuron chains

_____ 2. innervates cardiac muscle, smooth muscle, and exocrine glands

_____ 3. innervates skeletal muscle

_____ 4. consists of the axons of alpha motor neurons

_____ 5. exerts either an excitatory or inhibitory effect on its effector organs

_____ 6. dually innervates its effector organs

_____ 7. exerts only an excitatory effect on its effector organ

C. Matching

_____ 1. originates in the cranial-sacral region of the CNS		a. sympathetic nervous system
_____ 2. originates in the thoraco-lumbar region of the CNS		b. parasympathetic nervous system
_____ 3. dominates in "fight or flight" situations		
_____ 4. dominates in quiet, relaxed situations		

D. Fill-in-the-blank

The final common pathway in the control of motor activity is

____________________.

E. True/False

T/F 1. Atropine blocks all nicotinic receptor sites.

T/F 2. It is possible through the use of drugs to activate the receptors found in bronchiolar smooth muscle without influencing the receptors in the heart.

Neuromuscular Junction **(text page 209)**

Contents

Acetylcholine chemically links electrical activity in motor neurons with electrical activity in skeletal-muscle cells. p. 209

Acetylcholinesterase terminates acetylcholine activity at the neuromuscular junction. p. 211

A Closer Look at Exercise Physiology - Loss of Muscle Mass: A Plight of Space Flight p. 213

The neuromuscular junction is vulnerable to several chemical agents and diseases. p. 213

Section Synopsis

Each axon terminal of an alpha motor neuron forms a neuromuscular junction with a single muscle cell (fiber). Because these structures do not make direct contact, signals are passed between the nerve terminal and muscle fiber by means of the chemical messenger acetylcholine (ACh). An action potential in the axon terminal causes the influx of Ca^{++} into the axon terminal, which induces the release of ACh from its storage vesicles. The released ACh diffuses across the space separating the nerve and muscle cell and binds to special receptor sites on the underlying motor end-plate of the muscle-cell membrane. This combination triggers the opening of cation channels in the motor end-plate. The subsequent ion movements, predominantly Na^+ moving inward with lesser amounts of K^+ moving outward, depolarize the motor end-plate, producing the end-plate potential (EPP). Local current flow between the depolarized end plate and adjacent muscle-cell membrane brings these adjacent areas to threshold, initiating an action potential that is propagated throughout the muscle fiber. This muscle action potential triggers muscle contraction. Acetylcholinesterase inactivates ACh, terminating the EPP and, subsequently, the action potential. Disorders of neuromuscular transmission can occur as a result of interference with the release of ACh, blockage of the ACh receptor sites, or inhibition of acetylcholinesterase.

Learning Check (Answers on p. A-15)

A. Matching

_____1. binds with ACh receptor sites

_____2. causes explosive release of ACh

_____3. blocks release of ACh

_____4. inhibits acetylcholinesterase

_____5. antibodies inactivate ACh receptor sites

a. myasthenia gravis

b. black widow spider venom

c. curare

d. _Clostridium botulinum_ toxin

e. organophosphates

B. Indicate the proper sequence of events at the neuromuscular junction by filling in the blank with the appropriate number from 2 through 7. Numbers 1 and 8 are already identified.

_____ a. ACh is released from the axon terminal by exocytosis.

_____ b. An EPP takes place, primarily as a result of Na^+ influx.

__8__ c. Acetylcholinesterase inactivates ACh, terminating activity at the neuromuscular junction.

__1__ d. An action potential is propagated to an axon terminal of an alpha motor neuron.

_____ e. Channels that permit passage of Na^+ and K^+ are opened in the motor end-plate.

_____ f. Ca^{++} channels are opened in the axon terminal.

_____ g. Local current flow between the motor end-plate and adjacent muscle-cell membrane initiates an action potential that spreads throughout the muscle fiber.

_____ h. ACh binds with receptor sites on the motor end plate.

Chapter in Perspective **(text page 215)**

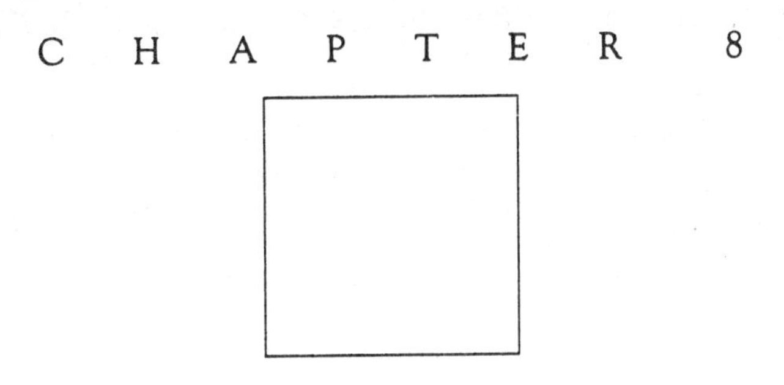

Muscle Physiology

Introduction (text page 216)

Structure of Skeletal Muscle (text page 217)

Contents

Skeletal-muscle fibers have a highly organized internal arrangement that creates a striated appearance. p. 217

Myosin forms the thick filaments whereas actin is the main structural component of the thin filaments. p. 220

Section Synopsis

Muscle cells are specialized for contraction. Through their contractile ability, they are able to shorten, develop tension, and perform work. Depending on their specialized structural and functional differences, there are three types of muscle: skeletal, smooth, and cardiac.

A single skeletal-muscle cell, known as a muscle fiber, is a long, cylindrical-shaped, multinucleated structure. Muscles are made up of bundles of muscle fibers wrapped in connective tissue. Muscle fibers are packed with myofibrils, which are specialized intracellular contractile elements. Each myofibril consists of alternating, slightly overlapping stacked sets of thick and thin filaments. This arrangement leads to a skeletal-muscle fiber's banding or striated microscopic appearance. Thick filaments, which are composed of the protein myosin, extend from one side to the other of each A (dark) band. Cross bridges made up of the myosin molecules' globular heads project from each thick filament toward the surrounding thin filaments. Each stack of thin filaments extends across an I (light) band and part way into the adjacent A bands so that the thin filaments partly overlap with the thick filaments. Thin filaments are composed primarily of the protein actin, which has the ability to bind and interact with the myosin cross-bridges in these overlapping regions to bring about contraction. However, two other proteins, tropomyosin and troponin, lie across the surface of the thin filament to prevent this cross-bridge interaction in the resting state. When a muscle fiber is stimulated to contract, calcium pulls these proteins from their blocking position to permit actin and myosin cross-bridge interaction so that contraction takes place.

The smallest contractile unit in a skeletal muscle fiber is a sarcomere, the segment of a myofibril that extends from one Z line to the next. The Z lines, which appear as dense vertical lines within the middle of each I band, consist of cytoskeletal proteins that join adjacent sarcomeres together.

Learning Check (Answers on p. A-16)

A. Matching

____ 1. dark band

____ 2. light band

____ 3. contains only thick filaments

____ 4. contains only thin filaments

____ 5. contains partially overlapping thick and thin filaments

____ 6. joins adjacent sarcomeres together

____ 7. runs down the middle of the A band

____ 8. runs down the middle of the I band

a. Z line

b. A band

c. I band

d. H zone

B. Using the following answer code, circle the appropriate letters to indicate which characteristic applies to each of the proteins within the muscle filaments:

a = myosin

b = actin

c = tropomyosin

d = troponin

1.	Forms the thick filaments	a	b	c	d
2.	Covers the cross bridge binding sites on actin	a	b	c	d
3.	Globular heads form cross bridge	a	b	c	d
4.	Spherical shaped molecules that form a helical chain	a	b	c	d
5.	Binds with Ca^{++}	a	b	c	d
6.	Found in thin filaments	a	b	c	d
7.	Holds tropomyosin in position	a	b	c	d
8.	Bind together at the cross bridges during muscle contraction	a	b	c	d
9.	Contractile proteins	a	b	c	d
10.	Regulatory proteins	a	b	c	d

C. The functional unit of skeletal muscle is a ____________.

Molecular Basis of Skeletal-Muscle Contraction (text page 222)

Contents

Section Synopsis

Excitation of a skeletal-muscle fiber by its motor neuron brings about contraction through a series of events that result in the thin filaments sliding closer together between the thick filaments, thus shortening each of the sarcomeres that compose the muscle fiber. This sliding-filament mechanism of muscle contraction is switched on by the release of Ca^{++} from intracellular storage depots in the lateral sacs of the sarcoplasmic reticulum. Calcium release occurs in response to propagation of the muscle-fiber action potential into the central portions of the fiber by means of the T tubules. Released Ca^{++} binds to the troponin-tropomyosin complex of the thin filament, pulling the complex slightly aside to uncover actin's cross-bridge binding sites. The exposed actin quickly attaches to a myosin cross-bridge that protrudes from the thick filament. The molecular interaction between actin and myosin releases the energy within the myosin head that was stored from the prior splitting of ATP by the myosin ATPase site. This released energy powers cross-bridge stroking. During a power stroke, an activated cross bridge bends toward the center of the thick filament, "rowing" in the thin filament to which it is attached. With the addition of a fresh ATP molecule to the myosin cross-bridge, myosin and actin detach and the cross bridge returns to its original conformation. The cross bridge then attaches to the actin molecule behind the one already moved forward and repeats the cycle. Repeated cycles of cross-bridge activity slide the thin filaments inward step by step. When there is no longer a local action potential, the lateral sacs actively take up the Ca^{++} and relaxation occurs. With Ca^{++} locked away once again, troponin and tropomyosin slip back into their blocking position, preventing further contact between actin and the myosin cross-bridges. The freed thin filaments slide back to their relaxed position. The entire contractile response lasts about one-hundred times longer than the action potential.

Learning Check (Answers on p. A-16)

A. Indicate what changes in banding pattern take place during contraction using the following answer code:

a = remains the same size during contraction

b = decreases in length (shortens) during contraction

____ 1. thick myofilament

____ 2. thin myofilament

____ 3. A band

____ 4. I band

____ 5. H zone

____ 6. sarcomere

B. What change in position of the filaments take place during contraction to account for these banding alterations?________

____

C. Matching

____ 1. Ca^{++}

____ 2. T tubules

____ 3. ATP

____ 4. lateral sacs of the sarcoplasmic reticulum

____ 5. myosin

____ 6. troponin-tropomyosin

____ 7. actin

a. cyclically binds with the myosin cross bridges during contraction

b. has ATPase activity

c. supplies energy for the power stroke of a cross bridge

d. rapidly transmits the action potential to the central portion of the muscle fiber

e. stores Ca^{++}

f. pulls the troponin-tropomyosin complex out of its blocking position

g. prevents actin from interacting with myosin when the muscle fiber is not excited

D. Which of the following is <u>not</u> involved in bringing about muscle relaxation?

a. Reuptake of Ca^{++} by the lateral sacs

b. No more ATP

c. No more action potential

d. Removal of ACh at the end plate by acetylcholinesterase

e. Filaments sliding back to their resting position

E. True/False

T/F Upon completion of an action potential in a muscle fiber, the contractile activity initiated by the action potential ceases.

Gradation of Muscle Contraction (text page 229)

Contents

Whole muscles are groups of muscle fibers bundled together by connective tissue and attached to bones by tendons. p. 229

The all-or-none law applies to individual muscle fibers but not to whole muscles. p. 229

The number of fibers contracting within a muscle depends on the extent of motor-unit recruitment. p. 230

The frequency of stimulation can influence the tension developed by each muscle fiber. p. 231

There is an optimal muscle length at which maximal tension can be developed upon a subsequent contraction. p. 232

Section Synopsis

Gradation of whole-muscle contraction can be accomplished by: (1) varying the number of muscle fibers contracting within the muscle; and (2) varying the tension developed by each contracting fiber. The greater the number of active muscle fibers, the greater the whole-muscle tension. The number of fibers contracting depends on: (1) the size of the muscle (the number of muscle fibers present); (2) the extent of motor-unit recruitment (how many motor neurons supplying the muscle are active); and (3) the size of each motor unit (how many muscle fibers are activated simultaneously by a single motor neuron).

Also, the greater the tension developed by each contracting fiber, the stronger the contraction of the whole muscle. Two readily variable factors having an effect on the fiber tension are: (1) the frequency of stimulation, which determines the extent of wave summation; and (2) the length of the fiber before the onset of contraction. Wave summation refers to the increase in external tension accompanying repetitive stimulation of the muscle fiber. After undergoing an action potential, the muscle cell membrane recovers from its refractory period and is able to be restimulated long before the Ca^{++} released during the excitation-contraction coupling process has been transported back into the lateral sacs. With sustained Ca^{++} release upon repetitive stimulation, the internal tension developed by cross-bridge activity lasts longer, permitting further stretching of the muscle fiber's noncontractile series elastic component (primarily the connective tissue). This increases the amount of external tension transmitted by the elastic component to the bone against a load. If the muscle fiber is stimulated so rapidly that it does not have a chance to start relaxing between stimuli, a smooth, sustained maximal contraction known as tetanus takes place - maximal for the fiber at that length.

The tension developed upon a tetanic contraction also depends on the length of the fiber at the onset of contraction. At the optimal length (l_0), which is the resting muscle length, there is maximal opportunity for cross-bridge interaction because of optimal overlap of thick and thin filaments; thus the greatest tension can be developed. At lengths shorter or longer than l_0, less tension can be developed upon contraction, primarily because of the reduced opportunity for participation by a portion of the cross bridges.

Learning Check (Answers on p. A-16)

A. Multiple Choice

1. A motor unit is :

   a. a whole muscle and all of the motor neurons innervating it

   b. a single muscle fiber and all of the motor neurons innervating it

   c. a single motor neuron and all of the muscle fibers it innervates

2. Which of the following is <u>not</u> a means by which gradation of skeletal muscle can be accomplished?

   a. recruiting variable numbers of motor units within a muscle

   b. stimulating a variable portion of each motor unit

   c. varying the frequency of stimulation

3. Which of the following statements concerning the length-tension relationship of skeletal muscle is <u>incorrect</u>?

   a. Maximum force can be produced at l_o.

   b. In the body the relaxed length of muscle is at its l_o.

   c. When a muscle is maximally stretched it can develop maximal tension upon contraction because the actin filaments can slide in a maximal distance.

   d. When the initial length of the muscle prior to contraction becomes very short, tension is decreased because of thin filament overlap; because the thick filaments are compressed against the Z lines; and because less Ca^{++} is released.

   e. When a muscle is longer or shorter than its optimal length, it will develop less than its maximum tension.

B. The contractile response of a muscle fiber to a single action potential is called a __________.

C. True/False

T/F The larger the motor units within a muscle, the more precisely controlled the gradations of contraction.

D. Use the answer code below to correctly complete the following explanation.

a = active state of the contractile elements

b = series elastic component

c = tetanus

d = wave summation

e = internal tension

f = external tension

Increased external tension developed upon repetitive action potentials is known as (1)__________. With repetitive stimulation, the period of time during which Ca^{++} is released and cross-bridge cycling is taking place, known as the (2)__________, is prolonged. This permits more time for the (3)__________ developed by cross-bridge cycling to further stretch the (4)__________, which transmits additional force to the bone as increased (5)__________. When a muscle fiber is stimulated so rapidly that no relaxation takes place between stimuli, a smooth, sustained maximal contraction known as (6)__________ takes place.

Metabolism and Types of Fibers (text page 235)

Contents

Section Synopsis

Several important steps in the muscle contractile response require ATP. Although little ATP itself is stored in muscle, there are three biochemical pathways available to furnish additional ATP: (1) ATP formation by the transfer of high-energy phosphates from stored creatine phosphate to ADP, providing the immediate source of additional ATP at the onset of exercise; (2) oxidative phosphorylation, which efficiently extracts large amounts of ATP from nutrient molecules if sufficient O_2 is available to support this system; and (3) glycolysis, which can synthesize ATP in the absence of O_2 but uses large amounts of stored glycogen and produces lactic acid in the process.

There are three types of muscle fibers based on which pathways they use for ATP synthesis (oxidative or glycolytic) and on the rapidity with which they split ATP and subsequently contract (slow twitch or fast twitch). The three fiber types are slow-oxidative (Type I) , fast-oxidative (Type IIa), and fast-glycolytic (Type IIb). Muscles consist of a mixture of these fiber types. A greater percentage of the oxidative fibers is found in muscles designed for low-intensity, long-duration work because these fibers are resistant to fatigue. Muscles adapted for short-duration, high-intensity activities contain a preponderance of the more powerful but fatigue-prone glycolytic fibers.

The exact cause of muscular fatigue is unclear, but lactic-acid accumulation with resultant inhibition of key muscle enzymes as well as depletion of energy stores are strong possibilities. Furthermore, even though the muscles themselves are capable of supporting further contractile activity, their response may be decreased by a reduction in stimulation because of neuromuscular or central fatigue.

To sustain contractile activity during exercise, an O_2 debt is frequently incurred that must be repaid after the exercise is over. Increased respiratory activity following exercise provides the O_2 required in the biochemical reactions that: (1) restore the creatine phosphate reserves; (2) remove lactic acid; and (3) replenish, at least in part, depleted glycogen stores, the remainder being accomplished in the long-term by food intake. Increased O_2 uptake after exercise is also due in part to other factors, such as elevated muscle temperature and increased levels of circulating epinephrine.

Although the distribution of muscle fiber types is genetically determined, the integrity and growth of muscle fibers, as well as their biochemical composition, are strongly influenced by their pattern of neural stimulation. Endurance (aerobic) exercise increases the oxidative capacity of oxidative fibers and may even convert fast-glycolytic fibers into fast-oxidative fibers. Power training, on the other hand, promotes enlargement of fast-glycolytic fibers, primarily by hypertrophy; increases their glycolytic capability; and transforms fast-oxidative fibers into fast-glycolytic fibers. In contrast, reduced activity of muscles, either through denervation or disuse, brings about their atrophy. Limited replacement of damaged muscle fibers is possible through repetition of a process used embryonically to form muscle.

Learning Check (Answers on p. A-17)

A. Fill-in-the-blank

1. The only energy source that can be used directly by the contractile machinery of a muscle fiber is __________.
2. The immediate source for supplying additional ATP at the onset of exercise is __________.
3. The two types of atrophy are __________ and____________.

B. Using the following answer code, indicate which of the characteristics in question apply to each of the muscle fiber types by circling the appropriate letter(s):

a = slow-oxidative fiber   b = fast-oxidative fiber

c = fast-glycolytic fiber

1. has high myosin ATPase activity   a b c
2. most resistant to fatigue   a b c
3. most readily fatigues   a b c
4. has numerous mitochondria   a b c
5. can be transformed into another fiber type by specific training   a b c
6. contains considerable myoglobin   a b c
7. found predominantly in muscles designed for endurance   a b c
8. has largest diameter   a b c
9. has abundant glycolytic enzymes   a b c
10. most powerful fibers   a b c
11. found predominantly in muscles adapted for short-duration, high intensity activities   a b c
12. produces the most lactic acid   a b c
13. hypertrophies in response to weight training   a b c
14. uses up considerable glycogen   a b c
15. requires a constant supply of O_2 a b c
16. has high yield of ATP for each nutrient molecule processed   a b c

Muscle Mechanics (text page 242)

Contents

The two primary types of contraction are isotonic and isometric. p. 242

The velocity of shortening is related to the load. p. 243

Although muscles can accomplish work, much of the energy is converted to heat. p. 243

Section Synopsis

The two primary types of muscle contraction - isometric (constant length) and isotonic (constant tension) - depend on the relationship between external muscle tension and the load. In both cases, internal tension is generated by cross-bridge cycling. If the subsequent external tension is less than the load, the muscle cannot shorten and lift the object but remains at constant length, producing an isometric contraction. In a concentric isotonic contraction, the external tension exceeds the load so the muscle can shorten and lift the object, maintaining constant tension throughout the period of shortening. In an eccentric isotonic contraction, the muscle contracts at constant tension in opposition to being passively stretched. Isometric contractions are important primarily for maintenance of posture, whereas isotonic contractions, along with other mixed types of contraction, accomplish movement.

The velocity of shortening during contraction is inversely related to the load, with velocity decreasing as load increases. Work accomplished by contraction of skeletal muscles depends on the weight of the object and the distance it is moved (work = force x distance). Isometric contractions accomplish no work; all of the energy used is converted into heat. About 25% of the energy consumed for an isotonic contraction is used for external work, the remainder being converted to heat.

Learning Check (Answers on p. A-17)

A. Matching

____1. Muscle tension exceeds the load

____2. Load exceeds muscle tension

____3. Length changes

____4. Length remains constant

____5. Tension remains constant

____6. Important in maintaining posture

____7. Used to accomplish movement

____8. Does not accomplish any work

a. isometric contraction

b. isotonic contraction

B. A(n) __________ contraction is an isotonic contraction in which the muscle shortens, whereas the muscle lengthens in a(n) __________ isotonic contraction.

C. True/False

T/F The velocity at which a muscle shortens is dependent entirely upon the ATPase activity of its fibers.

Control of Motor Movement (text page 244)

Contents

Section Synopsis

Control of any motor movement depends on the level of activity in the presynaptic inputs that converge on the alpha motor neurons supplying various muscles. These inputs come from three sources: (1) spinal-reflex pathways, which originate with afferent neurons; (2) the corticospinal descending system, which originates at the primary motor cortex and is concerned primarily with discrete, intricate movements of the hands; and (3) the multineuronal descending system, which originates in the brain stem and is mostly involved with postural adjustments and involuntary movements of the trunk and limbs. The final output from the motor regions of the brain stem is influenced by activity in many other regions of the brain, most notably the cerebellum, basal nuclei, and cerebral cortex. Imbalances between excitatory and inhibitory motor inputs arising from lesions in various regions of the motor systems can cause either flaccid or spastic paralysis.

Establishment and adjustment of motor commands depends on continuous afferent input, especially feedback about changes in muscle length (monitored by muscle spindles) and muscle tension (monitored by Golgi tendon organs). Activation of these receptor organs also gives rise to localized negative feedback reflexes that respectively resist changes in length and tension. Muscle spindles are located within the fleshy part of the muscle parallel to the muscle fibers. Muscle-spindle fibers are activated whenever the whole muscle is stretched. When alpha motor neurons trigger whole-muscle shortening, coactivation of gamma motor neurons that supply the muscle spindles occurs to take up the slack in the spindle, thereby maintaining its sensitivity to stretch. Golgi tendon organs are located within tendons. When contraction of muscle fibers pulls on the tendon, these receptors are stretched and activated in direct relation to the tension developed.

Learning Check (Answers on p. A-17)

A. Multiple choice

1. Which of the following provide direct input to alpha motor neurons?

   a. primary motor cortex
   b. brain stem
   c. cerebellum
   d. basal ganglia
   e. spinal reflex pathways

2. Disruption of a motor system that supplies inhibitory presynaptic inputs to motor neurons results in:

   a. flaccid paralysis
   b. uncoordinated, clumsy movements
   c. spastic paralysis
   d. paraplegia
   e. suppression of reflex activity

B. Fill-in-the-blank

1. Which descending motor pathway mediates performance of fine, discrete voluntary movements of the hands ?_______________

2. Which descending pathway is primarily concerned with regulation of posture involving involuntary movements of the trunk and limbs?__________________

3. _______________ motor neurons supply extrafusal muscle fibers whereas intrafusal fibers are innervated by _____________ motor neurons.

C. Indicate the characteristics of the muscle-tension receptors using the following answer code :

a = muscle spindle

b = Golgi tendon

c = both of these receptors

______ 1. monitors change in muscle length

______ 2. detects change in muscle tension

______ 3. activated by muscle stretch

______ 4. initiates a monosynaptic reflex when activated

______ 5. unless compensatory measures are taken, this receptor becomes slack

______ 6. involved in negative feedback

______ 7. provides information to motor regions of the brain

Smooth and Cardiac Muscle **(text page 249)**

Contents

Smooth and cardiac muscle share some basic properties with skeletal muscle. p. 249

Smooth muscle cells are small and unstriated. p. 249

Smooth-muscle cells are turned on by Ca^{++}-dependent phosphorylation of myosin. p. 249

Most groups of smooth-muscle tissue are capable of self-excitation. p. 252

The autonomic nervous system as well as other factors can modify the rate and strength of single unit smooth-muscle contraction. p. 253

Smooth muscle is a slow, economical contractile tissue. p. 254

Gradation of smooth-muscle contraction differs considerably from skeletal muscle. p. 254

Cardiac muscle blends features of both skeletal and smooth muscle. p. 256

Section Synopsis

The thick and thin filaments of smooth muscle are not arranged in an orderly pattern, so the fibers are not striated. Cytosolic Ca^{++}, which enters from the extracellular fluid as well as being released from sparse intracellular stores, activates cross-bridge cycling by initiating a series of biochemical reactions that result in phosphorylation of the myosin cross-bridges to enable them to bind with actin. The extent of cross-bridge activation depends on the concentration of cytosolic Ca^{++}, which in turn is influenced by a variety of inputs that affect smooth muscle contractile activity. Among these factors is the autonomic nervous system, which innervates smooth muscle. Multiunit smooth muscle is neurogenic, requiring stimulation of individual muscle fibers by this nerve supply to trigger contraction. Single-unit smooth muscle, on the other hand, is myogenic, being able to initiate its own contraction without any external influence as a result of spontaneous depolarizations to threshold potential brought about by automatic shifts in ionic fluxes. The autonomic nervous system, as well as other factors such as hormones and local metabolites, can modify the rate and strength of the self-induced contractions. Smooth-muscle contractions are energy efficient, enabling this type of muscle to economically sustain long-term contractions without fatigue. This, coupled with the fact that smooth muscle is able to exist at a variety of lengths with little change in tension, makes it ideally suited for its task of forming the walls of distensible hollow organs.

Cardiac muscle is found only in the heart. It has highly organized striated fibers like skeletal muscle. In common with single-unit smooth muscle, some cardiac muscle fibers are capable of generating action potentials, which are spread throughout the heart with the aid of gap junctions.

Learning Check (Answers on p. A-17)

A. Indicate which of the following characteristics are associated with each type of muscle by circling the appropriate letters using the answer code below:

a = characteristic of skeletal muscle

b = characteristic of multiunit smooth muscle

c = characteristic of single-unit smooth muscle

d = characteristic of cardiac muscle

1. Has T tubules and well-developed sarcoplasmic reticulum — a b c d
2. Ca^{++} enters from the ECF as well as being released from the sarcoplasmic reticulum — a b c d
3. Ca^{++}-dependent phosphorylation of myosin enables cross-bridge cycling — a b c d
4. Ca^{++} shifts the position of troponin and tropomyosin to enable cross-bridge cycling — a b c d
5. Contraction initiated by nervous stimulation (neurogenic) — a b c d
6. Self-excitable (myogenic). — a b c d
7. Abundant gap junctions. — a b c d
8. Striated. — a b c d
9. Contractile activity influenced by certain hormones. — a b c d

B. True/False

T/F 1. Pacemaker activity always initiates action potentials.

T/F 2. Slow wave potentials always initiate action potentials.

T/F 3. A single smooth muscle contractile response is much slower than a skeletal muscle twitch.

T/F 4. Smooth muscle is able to generate the same contractile tension per unit cross-sectional area as skeletal muscle but at considerably less energy expenditure.

T/F 5. All cross bridges are switched on by a single excitation in smooth muscle.

T/F 6. The greater the cytosolic Ca^{++} concentration in smooth muscle, the greater the tension developed.

T/F 7. Tension is developed in smooth muscle only in response to action potentials.

T/F 8. Smooth muscle can still develop tension even when considerably stretched because the thin filaments still overlap with the long thick filaments.

Chapter in Perspective (text page 256)

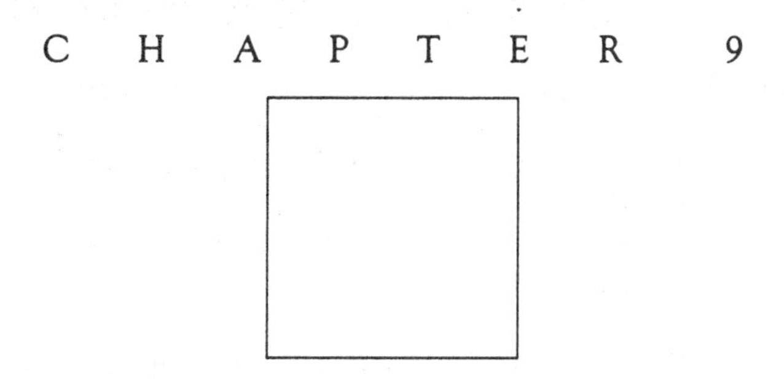

CARDIAC PHYSIOLOGY

Introduction (text page 258)

Anatomical Considerations (text page 259)

Contents

The heart is located in the middle of the chest cavity. p. 259

The heart is a dual pump. p. 260

Heart valves ensure the proper direction of blood flow through the heart. p. 262

The heart walls are composed primarily of spirally arranged cardiac-muscle fibers interconnected by intercalated discs. p. 265

The heart is enclosed by the pericardial sac. p. 266

Section Synopsis

The heart, which is enclosed by a fluid-filled lubricating sac that secures it in the middle of the chest cavity, is basically a dual pump that provides the driving pressure for blood flow through the pulmonary and systemic circulations. The heart has four chambers; there is an atrium, or venous input chamber, and a ventricle, or arterial output chamber, in each half of the heart. The heart valves direct the blood in the proper direction and prevent its flow in the reverse direction. The atria and ventricles are separated by a fibrous skeleton, which encircles and supports the four heart valves. The cardiac-muscle fibers are arranged spirally as a result of twisting of the embryonic heart on its axis. Contraction of these spirally arranged fibers produces a wringing effect important for efficient pumping. Also important for efficient pumping is the fact that the muscle fibers in each chamber act as a functional syncytium, contracting as a coordinated unit. The heart is self-excitable, initiating its own rhythmic contractions. Generation of energy for this contractile process places high oxygen demands on the cardiac muscle.

Learning Check (Answers on p. A-19)

A. Matching

_____ 1. receives low oxygenated blood from the venae cavae

_____ 2. prevents backflow of blood from the ventricles to the atria

_____ 3. pumps highly oxygenated blood into the aorta

_____ 4. prevents backflow of blood from arteries into ventricles

_____ 5. pumps low oxygenated blood into the pulmonary artery

_____ 6. receives highly oxygenated blood from the pulmonary vein

_____ 7. permits AV valves to function as one-way valves

a. AV valves

b. semilunar valves

c. left atrium

d. left ventricle

e. right atrium

f. right ventricle

g. chordae tendineae and papillary muscles

B. True/False

T/F 1. The left ventricle is a stronger pump than the right ventricle because it takes more blood to supply the body tissues than to supply the lungs.

T/F 2. The heart lies in the left half of the thoracic cavity.

T/F 3. Contraction of the spirally arranged cardiac muscle fibers produces a wringing effect for efficient pumping.

T/F 4. The atria and ventricles each act as a functional syncytium.

C. Fill-in-the-blank

1. Adjacent cardiac-muscle cells are joined end to end at specialized structures known as _______________, which contain two types of membrane junctions - _______________ and _______________.

2. The condition in which the pericardial sac becomes distended with excess fluid to the point that it impinges upon cardiac filling is known as _______________.

Electrical Activity of the Heart **(text page 266)**

Contents

The ECG is a record of the overall spread of electrical activity through the heart. p. 273

Various components of the ECG record can be correlated to specific cardiac events. p. 274

The ECG can be useful in diagnosing abnormal heart rates, arrhythmias, and myopathies. p. 275

<u>A Closer Look at Exercise Physiology - The What, Who, and When of Stress Testing</u> p. 278

Section Synopsis

The cardiac impulse originates at the SA node, the pacemaker of the heart, which has the fastest rate of spontaneous depolarization to threshold resulting from a cyclical reduction in passive K^+ efflux. Once initiated, the action potential spreads throughout the right and left atria, partially facilitated by specialized conduction pathways, but mostly by cell-to-cell spread of the impulse through gap junctions. The impulse passes from the atria into the ventricles through the AV node, the only point of electrical contact between these chambers. The action potential is delayed briefly at the AV node, assuring that atrial contraction precedes ventricular contraction to allow complete ventricular filling. The impulse then rapidly travels down the interventricular septum via the bundle of His and is rapidly dispersed throughout the myocardium by means of the Purkinje system. The remainder of the ventricular cells are activated by cell-to-cell spread of the impulse through gap junctions. Thus, the atria contract as a single unit, followed after a brief delay by a synchronized ventricular contraction.

The action potentials of contractile cardiac-muscle fibers exhibit a prolonged depolarization phase, or plateau, accompanied by a prolonged period of contraction, which assures adequate ejection time. This plateau is primarily due to activation of slow Ca^{++} channels. Because a long refractory period occurs in conjunction with this prolonged depolarization phase, summation and tetanus of cardiac muscle are impossible, thereby assuring the alternate periods of contraction and relaxation essential for pumping of blood.

The spread of electrical activity throughout the heart can be recorded from the surface of the body. This record, the ECG, can provide useful information about the status of the heart.

Learning Check (Answers on p. A-19)

A. Fill-in-the-blank

1. Ninety-nine percent of the cardiac fibers are specialized for _______________, whereas the remainder are specialized for _______________.

2. The slow drift of the membrane potential of pacemaker cells to threshold is believed to be caused by _______________.

3. The _______________ is the normal pacemaker of the heart.

4. The _______________ assures that atrial excitation and contraction are complete before ventricular excitation and contraction commence.

5. Tetanus of cardiac muscle is impossible because of _________________________________.

6 . ____________ is an abnormally slow heart rate whereas _____________is a rapid heart rate.

B. Matching

____ 1. uncoordinated excitation and contraction of the cardiac cells

____ 2. AV nodal damage

____ 3. overly irritable area that takes over pacemaker activity

a. heart block

b. ectopic focus

c. ventricular fibrillation

C. True/False

T/F 1. The only point of electrical contact between the atria and ventricles is the fibrous skeletal rings.

T/F 2. When ECF K^+ levels fall below normal, the heart becomes weak, flaccid, and dilated.

T/F 3. The ECG is an actual recording of cardiac electrical activity.

T/F 4. With 2:1 rhythm the atrial rate is very rapid and the ventricular rate is normal or above normal, whereas with 2:1 block the atrial rate is normal but the ventricular rate is below normal.

T/F 5. The plateau phase of the action potential in a contractile cardiac muscle cell occurs as a result of activation of slow Ca^{++} channels.

D. Which of the following is the proper sequence of cardiac excitation?

1. SA node → AV node → atrial myocardium → Bundle of His Purkinje fibers → ventricular myocardium.

2. SA node → atrial myocardium → AV node → Bundle of His ventricular myocardium → Purkinje fibers.

3. SA node → atrial myocardium → ventricular myocardium AV node → Bundle of His → Purkinje fibers.

4. SA node → atrial myocardium → AV node → Bundle of His Purkinje fibers → ventricular myocardium.

E. Matching

___ 1. ventricular depolarization

___ 2. time during which ventricles are contracting and emptying

___ 3. atrial depolarization

___ 4. time during which atria are repolarizing

___ 5. time during which impulse is traveling through the AV node

___ 6. time during which ventricles are filling

a. P wave

b. QRS complex

c. T wave

d. PR interval

e. TP interval

f. ST segment

Mechanical Events of the Cardiac Cycle (text page 278)

Contents

The heart alternately contracts to empty and relaxes to fill. p. 278

Two heart sounds associated with valve closures can be heard during the cardiac cycle. p. 282

Turbulent blood flow produces heart murmurs. p. 282

Section Synopsis

The cardiac cycle consists of three important events:

(1) the generation of electrical activity as the heart autorhythmically depolarizes and repolarizes, which can be recorded from the surface of the body by means of electrocardiography;

(2) mechanical activity consisting of alternate periods of systole (contraction and emptying) and diastole (relaxation and filling), which are initiated by the rhythmical electrical cycle; and

(3) directional flow of blood through the heart chambers, which is guided by valvular opening and closing induced by pressure changes that are generated by mechanical activity.

Valve closing gives rise to two normal heart sounds. The first heart sound is caused by closure of the atrioventricular (AV) valves, which signals the onset of ventricular systole. The second heart sound is due to closure of the aortic and pulmonary valves at the onset of diastole.

Reviewing the pressure changes associated with the cardiac cycle (see Figure 9-19), the atrial pressure curve remains low throughout the entire cardiac cycle, with only minor fluctuations occurring (normally varying between 0 to 8 mm Hg). The aortic pressure curve remains high the entire time, with moderate fluctuations (normally varying between a systolic pressure of 120 mm Hg to a diastolic pressure of 80 mm Hg). The ventricular pressure curve fluctuates dramatically because ventricular pressure must be below the low atrial pressure during diastole to allow the AV valves to open for filling to take place, and it must be above the high aortic pressure during systole to force the aortic valve open to allow emptying to occur. Therefore, ventricular pressure normally varies from 0 mm Hg during diastole to slightly more than 120 mm Hg during systole.

Defective valve function produces turbulent blood flow, which is audible as a heart murmur and can lead to detrimental circulatory consequences. Abnormal valves may be either stenotic and not open completely or insufficient and not close completely.

Learning Check (Answers on p. A-20)

A. Circle the correct choice to complete the statements

1. During ventricular filling ventricular pressure must be (greater than or less than) atrial pressure, while during ventricular ejection ventricular pressure must be (greater than or less than) aortic pressure. Atrial pressure is always (greater than or less than) aortic pressure. During isovolumetric ventricular contraction and relaxation, ventricular pressure is (greater than or less than) atrial pressure and (greater than or less than) aortic pressure.

2. A swishy murmur heard between the second and first heart sounds is indicative of a(n)(stenotic or insufficient) (AV or semilunar) valve.

3. The first heart sound is associated with closure of the (AV or semilunar) valves and signals the onset of (systole or diastole), whereas the second heart sound is associated with closure of the (AV or semilunar) valves and signals the onset of (systole or diastole).

Cardiac Output and Its Control (text page 283)

Contents

Section Synopsis

Cardiac output, the volume of blood ejected by each ventricle each minute, is determined by the heart rate times the stroke volume. Heart rate is varied by altering the balance of parasympathetic and sympathetic influence on the SA node, with parasympathetic stimulation slowing the heart rate and sympathetic stimulation speeding it up.

Stroke volume is dependent on: (1) the extent of ventricular filling, with an increased end-diastolic volume resulting in a larger stroke volume by means of the length-tension relationship (intrinsic control); (2) the contractility of the heart, which varies in relation to the degree of sympathetic stimulation (extrinsic control); and (3) the arterial blood pressure, which, if elevated, can reduce the effectiveness of the ventricles in ejecting blood.

Learning Check (Answers on p. A-20)

A. Indicate how the change listed in Column A affects the item listed in Column B by circling the appropriate letter.

Column A	increases	decreases	has no effect on	Column B
1. ↑heart rate	a	b	c	cardiac output
2. ↑stroke volume	a	b	c	cardiac output
3. exercise	a	b	c	cardiac output
4. ↑cardiac sympathetic activity	a	b	c	permeability of SA node to K^+
5. ↑cardia sympathetic activity	a	b	c	rate of depolarization of the SA node
6. ↑parasympathetic activity	a	b	c	rate of depolarization of the SA node
7. ↑parasympathetic activity	a	b	c	heart rate
8. ↑venous return	a	b	c	end-diastolic volume
9. ↑end-diastolic volume	a	b	c	stroke volume

COLUMN A	increases	decreases	has no effect on	COLUMN B
10. ↑length of cardiac muscle fiber prior to contraction	a	b	c	stroke volume
11. ↑venous return (under normal circumstances)	a	b	c	stroke volume
12. ↑venous return (with congestive heart failure)	a	b	c	stroke volume
13. ↑arterial blood pressure	a	b	c	stroke volume
14. ↑cardiac sympathetic activity	a	b	c	contractility of the ventricles
15. ↑parasympathetic activity	a	b	c	stroke volume
16. ↑parasympathetic activity	a	b	c	AV-nodal delay
17. ↑cardiac sympathetic activity	a	b	c	stroke volume
18. ↑afterload	a	b	c	time needed to pump out a normal stroke volume

COLUMN A	increases	decreases	has no effect on	COLUMN B
19. ↑parasympathetic activity	a	b	c	atrial contractility
20. ↑cardiac sympathetic activity	a	b	c	velocity of impulse conduction through the heart

Nourishing the Heart Muscle (text page 290)

Contents

The heart receives most of its own blood supply through the coronary circulation during diastole. p. 290

Atherosclerotic coronary artery disease can deprive the heart of essential oxygen. p. 291

The amount of cholesterol carried via high-density lipoproteins versus low-density lipoproteins is linked to atherosclerosis. p. 293

Section Synopsis

Cardiac muscle is supplied with oxygen and nutrients by blood delivered to it via the coronary circulation, not by blood within the heart chambers. Most coronary blood flow occurs during diastole, because the coronary vessels are compressed by the contracting heart muscle during systole. Coronary blood flow is varied to keep pace with cardiac oxygen needs. This is accomplished through coronary vasodilation mediated by adenosine, which is released in proportion to the oxygen demands of the heart muscle. In contrast to cardiac muscle's strong dependence on adequate oxygen delivery to maintain its aerobic metabolic activities, the heart is remarkably adaptable in its ability to use a variety of metabolic fuels, depending on their availability.

Coronary blood flow may be compromised by the development of atherosclerotic plaques, which can lead to ischemic heart disease ranging in severity from mild attacks of angina pectoris (chest pain) on exertion to fatal acute myocardial infarctions (heart attacks). The exact cause of atherosclerosis is unclear, but apparently the ratio of cholesterol carried in the plasma in conjunction with high-density lipoproteins (HDL) compared to low-density lipoproteins (LDL) is an important factor.

Learning Check (Answers on p. A-20)

A. True/False

T/F 1. Most blood flow through the coronary vessels occurs during ventricular systole when the heart is driving blood forward.

T/F 2. The heart utilizes glucose almost exclusively for energy production.

T/F 3. Cells obtain additional cholesterol by synthesizing LDL receptor proteins.

B. Fill-in-the-blank

1. The link that coordinates coronary blood flow with myocardial oxygen needs is believed to be ________________.

2. Insufficient circulation of oxygenated blood to the heart muscle to maintain aerobic metabolism is referred to as ________________.

3. Upon what two factors is the extent of myocardial infarction dependent? ________________ and ________________

4. List the four possible outcomes of an acute myocardial infarction.

________________ ________________

________________ ________________

5. ____________carries cholesterol to cells whereas____________ transports it away from cells.

6. The______________extracts cholesterol from the blood and converts in into______________, which are secreted into the bile.

7. The most accurate predictor of the risk of developing atherosclerosis is________________.

C. Matching

____ 1. consists of abnormal smooth muscle cells, cholesterol deposits, scar tissue, and possibly calcium deposits

____ 2. referred cardiac pain

____ 3. freely floating clot

____ 4. abnormal clot attached to a vessel wall

a. angina pectoris

b. embolus

c. atherosclerotic plaque

d. thrombus

Chapter in Perspective (text page 296)

C H A P T E R 1 0

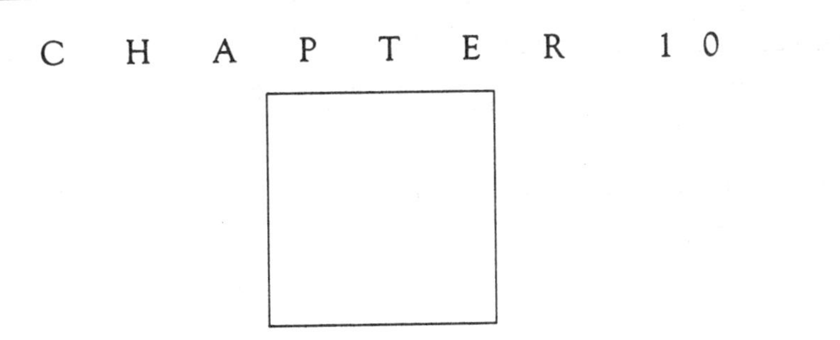

THE VASCULATURE AND BLOOD PRESSURE

Introduction (text page 298)

Contents

Materials are transported within the body by the blood as it is pumped through the blood vessels. p. 299

Blood flow through vessels is dependent on the pressure gradient and vascular resistance. p. 300

Section Synopsis

The vascular network interconnects various parts of the body with each other and with the external environment for the purpose of the exchange of materials. Organs that replenish nutrient supplies and remove metabolic wastes from the blood receive a disproportionately greater percentage of the cardiac output than is warranted by their metabolic needs. These "reconditioning" organs can better tolerate reductions in blood supply than can organs that receive blood solely for the purpose of meeting their own metabolic needs. The brain and heart are particularly vulnerable to reductions in blood supply. Therefore, the maintenance of adequate flow to these vulnerable organs is a high priority in circulatory function.

Blood flows in a closed loop between the heart and the tissues through a parallel arrangement of arteries. The arteries transport fresh blood throughout the body. The arterioles regulate the amount of blood that flows through each organ. The capillaries are the actual site of exchange of materials between the blood and surrounding tissue. The veins return the blood from the tissues to the heart.

The flow rate of blood through a vessel is directly proportional to the pressure gradient and inversely proportional to the resistance. The higher pressure at the beginning of a vessel is established by the pressure imparted to the blood by cardiac contraction. The lower pressure at the end is due to frictional losses as flowing blood rubs against the vessel wall. Resistance, the hindrance to blood flow through a vessel, is influenced most extensively by the vessel's radius. Resistance is inversely proportional to the fourth power of the radius, so small changes in resistance profoundly influence flow. As the radius increases, resistance decreases and flow increases. The converse is also true.

Learning Check (Answers on p. A-22)

A. True/False

T/F 1. Organs that recondition the blood normally receive considerably more of the cardiac output than is necessary to meet their metabolic needs.

T/F 2. Organs that do not adjust the blood are more vulnerable to reductions in blood flow than are the organs that perform homeostatic functions on the blood.

T/F 3. In general, each organ receives its own separate arterial blood supply because of the parallel arrangement of the vascular system.

B. Indicate what effect the changes in question would have on blood flow, using the following answer code :

a = This change would increase blood flow.

b = This change would decrease blood flow.

c = This change would have no effect on blood flow.

_____ 1. Increasing the pressure gradient in a vessel.
_____ 2. Increasing the resistance of a vessel.
_____ 3. Increasing the radius of a vessel.
_____ 4. Increasing the number of circulating red blood cells.
_____ 5. Increasing the length of a vessel.

Arteries (text page 302)

Contents

Section Synopsis

Arteries are large-radii, low-resistance passageways to the tissues and also serve as a pressure reservoir. Because of their elasticity, arteries expand to accommodate the extra volume of blood ejected into them by cardiac contraction and then recoil to continue driving the blood forward when the heart is relaxing.

Systolic pressure is the peak pressure exerted by the ejected blood against the vessel walls during cardiac systole. Diastolic pressure is the minimum pressure in the arteries when blood is draining off into the vessels downstream during cardiac diastole. Systolic and diastolic pressures can be indirectly measured using an inflatable external cuff whose pressure can be adjusted. The difference between systolic and diastolic pressure, the pulse pressure, can be felt as the pulse in the arteries that run near the surface of the body.

The average driving pressure throughout the cardiac cycle is the mean arterial pressure, which can be estimated using the following formula: mean arterial pressure = diastolic pressure + (1/3 x pulse pressure). Arterial pressure falls only slightly throughout the length of the arteries because these vessels offer little resistance to flow.

Learning Check (Answers on p. A-22)

A. List the two functions of arteries.

______________________________ and ______________________________

B. True/False

T/F 1. Flow of blood to the capillaries is intermittent in relationship to cardiac systole and diastole.

T/F 2. Arterial walls contain a thick layer of smooth muscle and an abundance of collagen and elastin fibers.

C. Assume a person has a blood pressure recording of 125/77:

1. What is the systolic pressure?______________________________

2. What is the diastolic pressure?______________________________

3. What is the pulse pressure?______________________________

4. What is the mean arterial pressure?______________________________

5. Would any sound be heard when the pressure in an external cuff around the arm was 130 mm Hg - yes or no? ________

6. Would any sound be heard when cuff pressure was 118 mm Hg?________

7. Would any sound be heard when cuff pressure was 75 mm Hg?________

Arterioles (text page 306)

Contents

Section Synopsis

To review, arterioles are the major resistance vessels. Their high resistance produces a large drop in mean pressure between the arteries and capillaries. This decline enhances blood flow by contributing to the pressure differential between the heart and the tissues. Tone, a baseline of contractile activity, is maintained in arterioles at all times. Arteriolar vasodilation, an expansion of arteriolar caliber above tonic level, decreases resistance and increases blood flow through the vessel, whereas vasoconstriction, a narrowing of the vessel, increases resistance and decreases flow.

Adjustments in arteriolar caliber can be accomplished independently in different tissues by various local control factors that change resistance to blood flow. Such adjustments are important in determining the distribution of cardiac output. Increased blood flow to an organ is accomplished by local arteriolar vasodilation, whereas less blood flows to organs that offer greater resistance as a result of arteriolar vasoconstriction. Increased metabolic activity within a tissue brings about chemical changes that induce arteriolar vasodilation, which increases blood flow to match the increased needs of the more active tissue, a response known as active hyperemia. Local metabolic influences are most important in skeletal muscle, heart, and brain arterioles. Similar chemical changes can be brought about by changes in a tissue's resting blood flow secondary to arterial occlusion or to alterations in mean arterial pressure. The resultant local influence on arteriolar smooth muscle adjusts resting flow toward normal. This effect is further enhanced by arteriolar smooth muscle's response to stretch; increased stretch results in myogenic vasoconstriction and decreased stretch to vasodilation. Other local factors come into play only under special circumstances. Release of histamine during injury and allergic reactions produces vasodilation in the insulted area, contributing to the swelling and redness of an inflammatory response. Arteriolar smooth muscle relaxes in response to local heat application, producing vasodilation, and constricts upon application of ice.

Arterioles are richly supplied by sympathetic nerve fibers, whose increased activity produces generalized vasoconstriction and a subsequent increase in mean arterial pressure. Decreased sympathetic activity, on the other hand, produces generalized arteriolar vasodilation, which lowers mean arteriolar pressure. These extrinsically controlled adjustments of arteriolar caliber help maintain the appropriate pressure head for driving blood forward to the tissues. Several hormones also extrinsically influence arteriolar resistance. The adrenal medullary hormones generally reinforce sympathetic activity. Vasopressin and angiotensin II, hormones primarily involved in fluid balance and maintenance of blood volume, also exert vasoconstrictor effects that are especially important during hemorrhage.

Learning Check (Answers on p. A-22)

A. Multiple Choice

1. Which of the following functions is (are) attributable to arterioles?

    a. responsible for a significant decline in mean pressure that helps establish the driving pressure gradient between the heart and tissues.

    b. site of exchange of materials between the blood and surrounding tissues.

    c. main determinant of total peripheral resistance.

    d. determine pattern of distribution of cardiac output

    e. play a role in the regulation of mean arterial blood pressure.

    f. convert the pulsatile nature of arterial blood pressure into a smooth, nonfluctuating pressure in the vessels further downstream.

    g. act as a pressure reservoir.

2. Which of the following tissues have powerful local control mechanisms?

   a. skin
   b. heart
   c. kidneys
   d. digestive organs
   e. skeletal muscle
   f. brain

B. Fill-in-the-blank

1. Arteriolar smooth muscle normally displays a state of partial constriction known as ____________________.

2. Relaxation of arteriolar smooth muscle causes the radius of the vessel to ____________________, a process known as ____________________, whereas contraction of arteriolar smooth muscle causes the lumen of the vessel to ____________________, a process known as ____________________.

3. What category of controls (local or extrinsic) is primarily responsible for matching tissue blood flow with the metabolic needs of the specific tissue involved? ____________________

4. What two hormones primarily involved in fluid balance are also potent vasoconstrictors? ____________________ and ____________________.

C. Indicate whether arteriolar vasoconstriction or vasodilation would occur in the tissue in question by circling the appropriate letter using the answer code.

a = would produce arteriolar vasoconstriction

b = would produce arteriolar vasodilation

c = would not cause any change in arteriolar caliber

1. decreased O_2 in skeletal muscle a b c

2. a hyperemic response in the heart a b c

3. histamine release in an injured tissue a b c

4. application of ice to a sprained ankle a b c

5. occlusion of an artery supplying a particular tissue. a b c

6. norepinephrine on cerebral arterioles a b c

7. sympathetic stimulation of kidney arterioles a b c

8. parasympathetic discharge on skeletal muscle arterioles a b c

9. increased stretch of arterioles supplying a particular tissue a b c

Capillaries (text page 313)

Capillaries are ideally suited to serve as sites of exchange. p. 313

Water-filled pores in the capillary wall permit passage of small, water-soluble substances that cannot cross the endothelial cells themselves. p. 314

Many capillaries are not open under resting conditions. p. 314

Diffusion across the capillary wall is important in solute exchange. p. 361

Bulk flow across the capillary wall is important in extracellular-fluid distribution. p. 317

The lymphatic system is an accessory route by which interstitial fluid can be returned to the blood. p. 321

Edema occurs when too much interstitial fluid accumulates. p. 323

Section Synopsis

The thin-walled, small-radius, extensively branched capillaries are ideally suited to serve as sites of exchange between the blood and surrounding tissues. Anatomically, the surface area for exchange is maximized and diffusion distance is minimized in the capillaries. Furthermore, because of the capillaries' large total cross-sectional area, the velocity of blood flow through them is relatively slow, which provides adequate time for exchanges to take place. The amount of blood flowing through a given capillary bed is dependent on the driving pressure gradient, the radii of the arterioles supplying the tissue, and the status of the precapillary sphincters that open or close a particular capillary in response to local tissue changes.

Two types of passive exchanges - diffusion and bulk flow - take place across capillary walls. Individual solutes are exchanged primarily by diffusion down concentration gradients. Lipid-soluble substances pass directly through the single layer of endothelial cells lining a capillary, whereas water-soluble substances pass through water-filled pores between the endothelial cells. Plasma proteins generally do not escape, except through some capillaries that are leakier (have larger pores) than others. Capillary permeability can be actively changed, as, for example, by the release of histamine, which widens the pores.

Imbalances in physical pressures acting across capillary walls are responsible for bulk flow of fluid through the pores back and forth between the plasma and interstitial fluid. According to the traditional view, fluid is forced out of the first portion of the capillary (ultrafiltration) where outward pressures (mainly capillary blood pressure) exceed inward pressures (mainly blood colloid osmotic pressure). Fluid is returned to the capillary along its last half when outward pressures fall below inward pressures. The reason for the shift in balance down the length of the capillary is the continuous decline in capillary blood pressure while the blood-colloid osmotic pressure remains constant. Bulk flow is responsible for the distribution of extracellular fluid between the plasma and interstitial fluid. The interstitial fluid can supply additional fluid to the plasma when needed to temporarily compensate for a reduced volume of circulating blood, or it can accommodate excess fluid that is shifted from the plasma to temporarily relieve an overexpanded blood volume.

Normally, slightly more fluid is filtered than is reabsorbed. The extra fluid, any leaked proteins, and tissue contaminants such as bacteria are picked up by the lymphatic system. Bacteria are destroyed as lymph passes through the lymph nodes en route to being returned to the venous system. A change in any of the pressures acting across the capillary membrane alters the normal slight imbalance between filtration and reabsorption. Edema results if excess interstitial fluid accumulates.

Learning Check (Answers on p. A-23)

A. True/False

T/F 1. Blood flows continuously through all capillaries.

T/F 2. Capillary walls consist of a single layer of endothelial cells.

T/F 3. The capillaries contain only 5% of the total blood volume at any point in time.

T/F 4. The same volume of blood passes through the capillaries per minute as passes through the aorta per minute, even though the velocity of blood flow is much slower in the capillaries.

T/F 5. Because there are no carrier transport systems in capillary walls, all capillaries are equally permeable.

T/F 6. Cells exchange material directly with the interstitial fluid, and the interstitial fluid, in turn, exchanges with the blood.

B. Question and answer.

1. What is the primary mechanism for exchange of individual solutes across the capillary wall?

2. What process is responsible for determining the distribution of the extracellular volume between the plasma and interstitial fluid?

3. What plasma constituent cannot readily pass through the capillary pores in most tissues?

4. What is responsible for the blood-colloid osmotic pressure?

5. What happens to the small excess quantity of fluid that is normally filtered but not reabsorbed?

C. Fill-in-the-blank

1. Movement of fluid out of the capillaries into the interstitial fluid is known as ______________________________.

2. Movement of fluid from the interstitial fluid into the plasma is known as ______________________________.

3. Accumulation of excess interstitial fluid is known as ______________________________.

D. Indicate what effect the change in question would have on bulk flow using the answer code below:

a = the change would increase ultrafiltration and decrease reabsorption

b = the change would decrease ultrafiltration and increase reabsorption

c = the change would increase both ultrafiltration and reabsorption

d = the change would decrease both ultrafiltration and reabsorption

e = the change would not alter ultrafiltration and reabsorption

__________1. loss of plasma protein in the urine due to kidney disease.

__________2. rise in capillary blood pressure in connection with congestive heart failure.

__________3. loss of plasma volume due to hemorrhage

__________4. escape of plasma proteins into the interstitial fluid due to capillary damage

__________5. reduced synthesis of plasma proteins due to liver disease

__________6. expanded plasma volume due to excessive fluid intake

Veins (text page 324)

Contents

Veins serve as a blood reservoir as well as passageways back to the heart. p. 324

Venous return is enhanced by a number of extrinsic factors. p. 325

Section Synopsis

Veins are large radius, low-resistance passageways for return of blood from the tissues to the heart. Additionally, they can accommodate variable volumes of blood, thus acting as as blood reservoir. The capacity of veins to hold blood can change markedly with little change in venous pressure. Veins are thin-walled, highly distensible vessels that can passively stretch to store a larger volume of blood. Venous capacity is reduced by sympathetically induced venous vasoconstriction and by external compression of the veins through contraction of surrounding skeletal muscles. Both of these factors drive blood out of the veins and increase venous return to the heart. Both are also important in counteracting the effect of gravity, which tends to cause pooling of blood in the leg veins when a person is in an upright position. Such pooling reduces venous return and lowers the effective circulating volume. The skeletal-muscle pump is especially important in pushing the pooled blood forward. One-way venous valves assure that blood is driven toward the heart and prevented from flowing back toward the tissues.

The primary force responsible for venous flow is the pressure gradient between the veins and atrium (that is, what remains of the driving pressure imparted to the blood by cardiac contraction.) Venous return is also enhanced by the respiratory pump and cardiac-suction effect. Respiratory activity produces a subatmospheric pressure in the thoracic cavity, which establishes an external pressure gradient that encourages flow from the lower veins to the chest veins that empty into the heart. In addition, slightly negative pressures created within the atria during ventricular systole and within the ventricles during ventricular diastole exert a suctioning effect that further enhances venous return and facilitates cardiac filling.

Learning Check (Answers on p. A-23)

A. Which of the following characteristics apply to veins?

1. Low-resistance vessels
2. Slow velocity of blood flow
3. Highly distensible
4. Display elastic recoil
5. Innervated by sympathetic nerves
6. Have little myogenic tone
7. Serve as passageways from tissues to heart
8. Act as a pressure reservoir
9. Act as a blood reservoir
10. Contain one-way valves
11. Resistance increased upon vasoconstriction
12. Normally contain 60% of the total blood volume

B. Indicate whether the following factors increase or decrease venous return using the following answer code:

a = increases venous return

b = decreases venous return

c = has no effect on venous return

______1. sympathetically-induced venous vasoconstriction

______2. skeletal muscle activity

______3. gravitational effects on the venous system

______4. respiratory activity

______5. increased atrial pressure associated with a leaky AV valve.

______6. ventricular pressure change associated with diastolic recoil

C. True/False

T/F 1. Cardiac contraction induces blood flow in the arterial system but it has no influence on blood flow in the veins.

T/F 2. As a result of gravitational effects, venous pressure in the lower extremities is greater when a person is standing up than when lying down.

T/F 3. Edema, venous pooling, and a reduction in cardiac output are consequences of standing still for a long time.

T/F 4. The effective circulating volume is reduced when blood collects in distended varicose veins.

T/F 5. It is important to try to hold a fainted person upright.

Blood Pressure (text page 329)

Contents

Section Synopsis

Regulation of mean arterial pressure depends on control of its two main determinants, cardiac output and total peripheral resistance. Control of cardiac output, in turn, depends on regulation of heart rate and stroke volume, whereas total peripheral resistance is determined primarily by the degree of arteriolar vasoconstriction. Short-term regulation of blood pressure is accomplished primarily by the baroreceptor reflex. Carotid-sinus and aortic-arch baroreceptors continuously monitor mean arterial pressure and pulse pressure. When they detect a deviation from normal, they signal the medullary cardiovascular center, which responds by adjusting autonomic output to the heart and blood vessels to restore the blood pressure to normal. Long-term control of blood pressure involves maintenance of proper plasma volume through the kidneys' control of salt and water balance. Occasionally, other reflexes and responses directed primarily toward another goal exert an overriding effect on the normal blood pressure maintenance systems - for example, during exercise, fight or flight responses, and temperature regulatory mechanisms.

Blood pressure can be abnormally high (hypertension) or abnormally low (hypotension). The cause of primary hypertension, which represents 90% of the cases, is unknown but apparently involves some abnormality in salt management in the body. In the remaining 10% of the cases, hypertension is secondary to a known disease involving another body system, either cardiovascular, renal, endocrine, or neural. The baroreceptors' sensitivity is reduced in the presence of sustained hypertension so that these receptors regulate blood pressure (that is, prevent fluctuations in blood pressure) at the new higher level.

Transient hypotension occurs as a result of inadequate autonomic activity when a person first stands up following prolonged bed rest (orthostatic hypotension) or as a result of inappropriate autonomic activity as in the loss of sympathetic vasoconstrictor tone that sometimes accompanies severe emotional stress responses (emotional fainting). Severe sustained hypotension resulting in generalized inadequate blood delivery to the tissues is known as circulatory shock. It may result from loss of plasma volume, failure of the heart, widespread arteriolar vasodilation induced by toxic or allergic vasodilator substances, or neurally defective vasoconstrictor tone. Compensatory measures may be able to restore blood pressure to adequate levels, but sometimes irreversible shock occurs if the cardiovascular system itself deteriorates as a result of side-effects of the compensatory responses.

Learning Check (Answers on p. A-24)

A. Fill-in-the-blank

1. Mean arterial pressure = ________________ X ________________.

2. Cardiac output = ________________ X ________________.

3. Total peripheral resistance is primarily dependent on ______________________________.

4. The receptors for the baroreceptor reflex are located in the ________________ and ________________, whereas the integrating center is the ________________ located in ______________________________. The efferent pathway for this reflex is the ________________ The effector organs are the ________________ and ________________.

B. Indicate what compensatory changes occur in the factors in question to restore the blood pressure to normal in response to hypovolemic hypotension resulting from severe hemorrhage.

a = increased

b = decreased

c = no effect

_____1. rate of afferent firing generated by the carotid sinus and aortic arch baroreceptors

_____2. sympathetic output by the cardiovascular center

_____3. parasympathetic output by the cardiovascular center

_____4. heart rate

_____5. stroke volume

_____6. cardiac output

_____7. arteriolar radii

_____8. total peripheral resistance

_____9. venous radii

____10. venous return

____11. urinary output

____12. fluid retention within the body

____13. fluid movement from the interstitial fluid into the plasma across the capillaries

C. True/False

T/F 1. Cardiovascular changes associated with exercise can be accounted for by local metabolic influences on skeletal muscle arterioles coupled with adjustments via the baroreceptor reflex.

T/F 2. Primary hypertension refers to chronically elevated blood pressure of unknown origin.

T/F 3. The baroreceptors are usually no longer functional in the presence of hypertension.

T/F 4. Fainting may occur as a result of inappropriate loss of vascular tone in response to highly emotional situations.

T/F 5. Irreversible shock occurs when the cardiovascular system itself starts to deteriorate as a consequence of side-effects from compensatory measures to the severe hypotension.

D. Multiple choice

1. Which of the following may cause secondary hypertension?
   a. atherosclerosis
   b. diabetes mellitus
   c. occlusion of a renal artery
   d. pheochromocytoma
   e. AV valve stenosis
   f. Conn's syndrome
   g. brain tumor

2. Which of the following may cause circulatory shock?
   a. severe hemorrhage
   b. extensive sweating
   c. severe diarrhea or vomiting
   d. generalized infections
   e. generalized allergic reactions
   f. heart failure
   g. severe crushing injuries

Chapter in Perspective (text page 340)

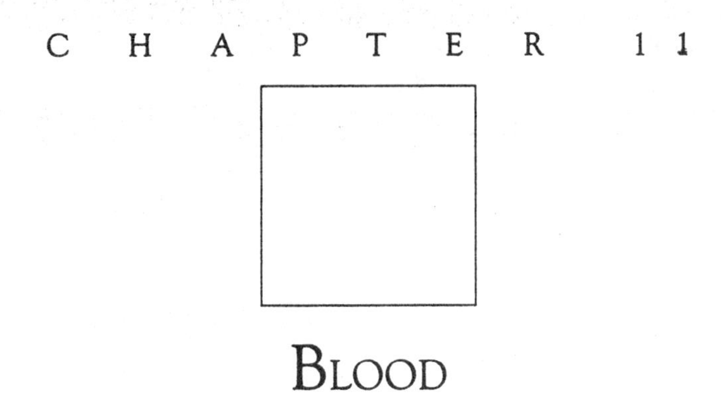

C H A P T E R 11

BLOOD

Introduction (text page 342)

Plasma (text page 343)

Contents

Many of the functions of plasma are carried out by plasma proteins. p. 343

Section Synopsis

The 5 to 5.5 liter volume of blood in an adult consists of 42-45% erythrocytes, less than 1% leukocytes and platelets, and 55-58% plasma. The percentage of whole blood volume occupied by erythrocytes is known as the hematocrit.

Plasma is a complex liquid that serves as a transport medium for substances being carried in the blood. Also, because of its high heat capacity, plasma absorbs, distributes, and carries for elimination to the environment the heat generated by metabolic activity in the tissues. The inorganic constituents in plasma are important in membrane excitability, osmotic distribution of fluid between the ECF and ICF, and buffering of pH changes. All plasma constituents except the plasma proteins, the largest and most abundant of organic constituents, are freely diffusible across the capillary walls.

The three classes of plasma proteins - albumins, globulins, and fibrinogens - perform a variety of functions, the most important of which are: (1) retention of fluid in the vasculature; (2) acid-base buffering of the blood; (3) transport of particular plasma constituents such as hormones; (4) defense against foreign invaders; and (5) blood clotting.

Learning Check (Answers on p. A-25)

A. Fill-in-the-blank

1. The percentage of whole blood occupied by erythrocytes, known as the __________, is normally (greater than, less than) the plasma volume.

2. The three types of cellular elements in the blood are _____________, _____________, and _____________.

3. By far the most abundant of these cell types are the _____________.

B. True/False

T/F 1. Blood can absorb metabolic heat while inducing only small changes in the temperature of the blood itself.

T/F 2. All of the constituents present in the plasma are freely diffusible across the capillary walls.

C. Which of the following is <u>not</u> a function served by the plasma proteins?

1. facilitate retention of fluid in the vasculature
2. important in blood clotting
3. bind and transport certain hormones in the blood
4. transport O_2 in the blood
5. serve as antibodies
6. contribute to buffering capacity of the blood

Erythrocytes (text page 344)

Contents

Section Synopsis

Erythrocytes are specialized for transport of O_2 and CO_2 in the blood. Their unique biconcave disc shape facilitates the diffusion of these gases into and out of the red blood cells, whereas their flexible plasma membrane permits them to squeeze through the narrowest of capillaries. They do not contain a nucleus, organelles, or ribosomes but instead are packed full of hemoglobin. Hemoglobin is an iron-containing molecule that can loosely, reversibly bind with O_2, CO_2 and H^+, making it indispensible for O_2 and CO_2 transport as well as being important as a buffer. The erythrocyte cytosol contains several nonrenewable enzymes: (1) glycolytic enzymes, which anaerobically direct ATP formation, and (2) carbonic anhydrase, which catalyzes the ultimate conversion of CO_2 into HCO_3^-, the primary form in which CO_2 is transported in the blood.

Unable to replace cell components, erythrocytes are destined to a short life span of about 120 days, by which time the plasma membrane is usually so fragile that it ruptures as the cell squeezes through the tight capillaries of the spleen. Macrophages engulf and degrade the cellular debris, which is recycled except for a portion of the heme molecule that is converted into bilirubin and secreted in the bile. Undifferentiated stem cells in the red bone marrow give rise to all cellular elements of the blood. Erythrocyte production (erythropoiesis) by the marrow normally keeps pace with the rate of erythrocyte destruction or loss to keep the red cell count constant. Erythropoiesis is stimulated by erythropoietin, a hormone secreted by the kidneys in response to reduced O_2 delivery.

A reduced O_2-carrying capacity of the blood, anemia, can occur as a consequence of a deficiency of essential ingredients such as iron, folic acid, or vitamin B_{12}; failure of the erythropoietic capabilities of the bone marrow; lack of sufficient marrow stimulation arising from a deficiency of erythropoietin; or excessive loss of red blood cells through hemorrhage or hemolysis. On the other hand, polycythemia, an excess of erythrocytes, can occur as a result of their overproduction by a tumorlike condition of the bone marrow or as a compensatory response to prolonged reduction in O_2 delivery to the tissues.

Learning Check (Answers on p. A-25)

A. True/False

T/F 1. Hemoglobin can only carry O_2.

B. Fill-in-the-blank

1. Erythrocytes contain the enzyme __________, which catalyzes the conversion of metabolically-produced CO_2 into HCO_3.
2. Most of the old erythrocytes are removed from the circulation as they rupture passing through the narrow capillaries of the __________.
3. Red blood cell production by the __________ is stimulated by the hormone __________ , which is secreted from the kidney into the blood in response to _____________________.

C. Matching

____1. deficiency of intrinsic factor

____2. insufficiency of iron to synthesize adequate hemoglobin

____3. destruction of bone marrow

____4. abnormal loss of blood

____5. tumorlike condition of bone marrow

____6. inadequate erythropoietin secretion

____7. excessive rupture of circulating erythrocytes

____8. associated with living at high altitudes

a. hemolytic anemia

b. aplastic anemia

c. iron deficiency anemia

d. hemorrhagic anemia

e. pernicious anemia

f. renal anemia

g. primary polycythemia

h. secondary polycythemia

D. Multiple Choice

1. Which of the following statements concerning iron is (are) <u>incorrect</u>?

   a. Iron is found in the heme portion of the hemoglobin molecule.

   b. Iron is converted into bilirubin upon erythrocyte degradation.

   c. Iron readily combines reversibly with O_2.

   d. Diets deficient in iron can lead to anemia.

   e. Iron is not adequately absorbed from the digestive tract in pernicious anemia.

2. Which of the following statements concerning erythrocytes is (are) incorrect?

   a. Erythrocytes do not contain a nucleus, organelles, or ribosomes.

   b. Erythrocytes are always shaped like biconcave discs.

   c. Erythrocytes originate from the same undifferentiated stem cells as leukocytes and platelets.

   d. Erythrocytes are unable to utilize the O_2 they contain for their own ATP formation.

   e. Erythrocytes only live about 120 days.

Leukocytes (text page 350)

Contents

Leukocytes function primarily outside of the blood. p. 350

There are five different types of leukocytes. p. 351

Leukocytes are produced at varying rates depending on the changing defense needs of the body. p. 351

Section Synopsis

Leukocytes (white blood cells) are the defense corps of the body. They attack foreign invaders, destroy abnormal cells that arise in the body, and clean up cellular debris. There are five types of leukocytes, each with a different task. (1) Neutrophils, the phagocytic specialists, are important in engulfing bacteria and debris. (2) Eosinophils specialize in attacking parasitic worms and play a key role in allergic responses. (3) Basophils release two chemicals: histamine, which is also important in allergic manifestations, and heparin, which helps clear fat particles from the blood and may play a role as an anticoagulant. (4) Monocytes, upon leaving the blood, set up residence in the tissues and greatly enlarge to become the large tissue phagocytes known as macrophages. (5) Lymphocytes are primarily responsible for the specific immune defenses of the body. B lymphocytes produce antibodies, whereas T lymphocytes participate in cell-mediated immune responses.

Leukocytes are present in the blood only while in transit from their site of production and storage in the bone marrow (and also in the lymphoid organs in the case of the lymphocytes) to their site of action in the tissues. Accordingly, at any given time only a small percentage of the total leukocyte population is in the blood. The majority of the leukocytes are in storage or out in the tissues on surveillance missions or performing actual combative activities. All leukocytes have a limited life span and must be replenished by ongoing differentiation and proliferation of precursor cells. The total number and percentage of each of the different types of leukocytes produced varies depending on the defense needs of the body. The control mechanisms involved in maintaining and adjusting leukocyte production are only beginning to be unraveled.

Ironically, the body's defense capabilities are devastatingly reduced both when the bone marrow produces too few leukocytes, as with toxic damage of the marrow, and when too many white blood cells are produced, as in leukemia. The uncontrolled proliferation of leukocytes accompanying this cancerous condition results in the release of abnormal or immature cells that are unable to participate in normal defense activities.

Learning Check (Answers on p. A-26)

A. Fill-in-the-blank

1. In what two different ways do leukocytes defend against foreign invasion by infectious agents? ________________ and ____

2. Which type of leukocyte is produced primarily in lymphoid tissue? ___________________

3. Which of the types of leukocytes are categorized as polymorphonuclear granulocytes? ____________________ ____

B. True/False

T/F 1. White blood cells spend the majority of their time in the blood.

T/F 2. Defense capabilities of the body are reduced in leukemia even though there are an excessive number of leukocytes.

C. Matching

____1. most abundant type of granulocyte

____2. become tissue macrophages

____3. produce antibodies

____4. first phagocytes to arrive at site of bacterial invasion

____5. release histamine and heparin

____6. destroy parasitic worms

____7. participate in cell-mediated immune responses

____8. most abundant type of agranulocyte

____9. similar to mast cells

a. neutrophils

b. eosinophils

c. basophils

d. monocytes

e. lymphocytes

Platelets and Hemostasis (text page 354)

Contents

Inappropriate clotting is responsible for thromboembolism. p. 360

Hemophilia is the primary condition responsible for excessive bleeding. p. 360

Section Synopsis

Platelets are cell fragments derived from large megakaryocytes in the bone marrow. They play an important role in hemostasis, the arrest of bleeding from an injured vessel. The three main steps in hemostasis are: (1) vascular spasm, (2) platelet plugging, and (3) clot formation. Vascular spasm reduces blood flow through an injured vessel, whereas aggregation of platelets at the site of vessel injury quickly plugs the defect. Platelets start to aggregate upon contact with exposed collagen in the damaged vessel wall. The platelet plug continues to build as a result of the release of ADP and thromboxane A_2 from the aggregated platelets, which act in positive-feedback fashion to cause more platelets to pile on. Other chemicals released from the platelet plug induce secondary vasoconstriction and are essential for clot formation. Platelet accumulation is restricted to the damaged area because of the release of an inhibitor of platelet aggregation, prostacyclin, from adjacent healthy vascular tissue.

Clot formation (blood coagulation) reinforces the platelet plug and converts blood in the vicinity of a vessel injury into a non-flowing gel. The majority of factors necessary for clotting are always present in the plasma in inactive precursor form. When a vessel is damaged, exposed collagen initiates a cascade of reactions involving successive activation of these clotting factors, ultimately converting fibrinogen into fibrin. Fibrin, an insoluble thread-like molecule, is laid down as the meshwork of the clot, which entangles blood cells to complete clot formation. Retraction of the clot subsequently pulls the wound edges closer together, extruding the excess fluid, which is known as serum. Blood that has escaped into the tissues is also coagulated upon exposure to tissue thromboplastin, which likewise sets the clotting process into motion but shortcuts the longer pathway involved in intravascular clotting.

When no longer needed, clots are dissolved by plasmin, a fibrinolytic factor also activated by exposed collagen. Low level plasmin activity normally goes on continuously to keep the vascular highways cleared of the small amounts of fibrin that are constantly being formed. The role of other naturally occurring anticoagulants is unclear.

Disorders of hemostasis include inappropriate clot formation in intact vessels (thromboembolism) and excessive bleeding resulting from a deficiency of one of the clotting factors (hemophilia) or to a reduction in platelets (thrombocytopenia purpura).

Learning Check (Answers on p. A-26)

A. True/False

T/F 1. Being a cell fragment, a platelet lacks a nucleus and organelles and does not have any synthetic ability.

T/F 2. Hemostatic mechanisms are effective for completely arresting bleeding from small vessels, but rupture of medium to large vessels requires external intervention to stop bleeding.

T/F 3. Serum is identical to plasma.

B. Fill-in-the-blank

1. List the three major steps in hemostasis:

1______________________________

2______________________________

3______________________________

2. The majority of plasma clotting factors are synthesized by the ____________________.

C. Which of the following is <u>not</u> triggered by exposed collagen?

a. vascular spasm

b. platelet aggregation

c. activation of the clotting cascade

d. activation of plasminogen

D. Matching

_____ 1. causes platelets to aggregate in positive-feedback fashion

_____ 2. activates prothrombin

_____ 3. fibrinolytic enzyme

_____ 4. inhibits platelet aggregation

_____ 5. first factor activated in intrinsic clotting pathway

_____ 6. forms the meshwork of the clot

_____ 7. stabilizes the clot

_____ 8. activates fibrinogen

_____ 9. activated by tissue thromboplastin

_____10. factor deficient in Hemophilia A

a. prostacyclin

b. Factor VIII

c. plasmin

d. ADP

e. fibrin

f. thrombin

g. Factor X

h. Factor XII

i. Factor XIII

Chapter in Perspective (text page 361)

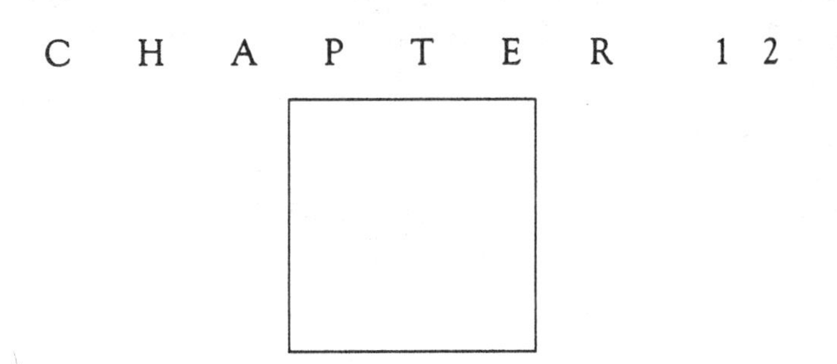

CHAPTER 12

DEFENSE MECHANISMS OF THE BODY

Introduction (text page 362)

Contents

The immune defense system provides protection against foreign and abnormal cells and removes cellular debris. p. 363

Pathogenic bacteria and viruses are the major targets of the immune defense system. p. 363

Leukocytes are the effector cells of the immune defense system. p. 363

Immune responses may be either specific or nonspecific. p. 364

Section Synopsis

Foreign invaders and newly arisen mutant cells are immediately confronted with multiple interrelated defense mechanisms aimed at destroying and eliminating anything that is not part of normal self. These mechanisms, collectively referred to as immunity, include both nonspecific and specific immune responses. Nonspecific responses, which nonselectively defend against foreign material even upon initial exposure to it, include: (1) inflammation mediated largely by neutrophils and monocyte-derived macrophages; (2) interferon; (3) natural killer cells; and (4) the complement system. Specific responses, which are aimed at the destruction of particular invaders to which the body has had prior exposure and is specially prepared for selective attack, are accomplished by lymphocytes. Thus, leukocytes and their derivatives are the major effector cells of the immune system, being reinforced by a number of different plasma proteins. Leukocytes are produced in the bone marrow, then circulate transiently in the blood. They spend most of their time, however, on defense missions in the tissues. Some lymphocytes are also produced, differentiated, and perform their defense activities within lymphoid tissues strategically located at likely points of foreign infiltration.

The most common invaders are bacteria and viruses. Bacteria are self-sustaining, single-celled organisms, which produce disease by virtue of the destructive chemicals they release. Viruses are protein-coated nucleic acid particles, which invade host cells and take over the cellular metabolic machinery for their own survival to the detriment of the host cell. In addition to defending against these microbes and mutant cells, the immune cells also clean up cellular debris, preparing the way for tissue repair.

Learning Check (Answers on p. A-27)

A. Multiple choice

1. Which of the following is <u>not</u> attributable to the immune defense system?

    a. Defends against pathogenic microorganisms

    b. Converts foreign chemicals into compounds that can be more readily eliminated in the urine

    c. Removes worn-out cells and tissue debris

    d. Identifies and destroys abnormal or mutant cells

    e. Can inappropriately induce allergic responses and autoimmune diseases

    f. Is responsible for rejection of transplanted organs

2. Which of the following statements concerning leukocytes is (are) <u>incorrect</u>?

   a. Monocytes are transformed into macrophages.

   b. T lymphocytes are transformed into plasma cells that secrete antibodies.

   c. Neutrophils are highly mobile phagocytic specialists.

   d. Basophils release histamine.

   e. Lymphocytes arise in part from lymphoid tissues.

B. Circle the appropriate answer using the following code to differentiate between bacterial and viral characteristics:

a = pertains to bacteria

b = pertains to viruses

1. consists of nucleic acids enclosed by a protein coat — a b

2. self-sustaining, single-celled organisms — a b

3. can induce host cells to produce substances toxic to the cell — a b

4. can secrete enzymes or toxins that are injurious to host cells — a b

5. can transform normal host cells into cancer cells — a b

6. can become incorporated into a host cell such that the body's defense mechanisms turn against the cell — a b

C. True/False

T/F 1. The spleen clears the lymph that passes through it of bacteria and other foreign matter.

T/F 2. Specific immune responses are selectively targeted against particular foreign material to which the body has previously been exposed.

T/F 3. The complement system can only be activated by antibodies.

T/F 4. Specific immune responses are accomplished by neutrophils.

D. Matching

_____1. a family of proteins that nonspecifically defend against viral infection

_____2. response to tissue injury of any origin in which neutrophils and macrophages play a major role

_____3. a group of inactive plasma proteins which when activated bring about destruction of foreign cells by attacking their plasma membranes

_____4. lymphocytelike entities, which spontaneously lyse a variety of tumor cells and virus-infected host cells

a. complement system

b. natural killer cells

c. interferon

d. inflammation

Nonspecific Immune Responses (text page 365)

Contents

Section Synopsis

Nonspecific immune responses form a first line of defense against the presence of atypical cells (foreign, mutant, or injured cells) even upon first exposure to them. Inflammation, interferon, natural killer cells, and the complement system all contribute nonselectively to defense of the body. Inflammation is a nonspecific response to foreign invasion or tissue damage mediated largely by the professional phagocytes (neutrophils and monocytes-turned-macrophages) and their secretions. Resident tissue macrophages immediately start the battle locally at the affected site. Meanwhile, histamine-induced vasodilation and increased permeability of local vessels permit enhanced delivery of leukocytes and inactive plasma protein precursors crucial to the inflammatory process. These vascular changes are also largely responsible for the observable local manifestations of inflammation - namely swelling, redness, heat, and pain.

Upon activation by locally-released factors, the following plasma proteins participate in the inflammatory process: (1) clotting factors, which form fibrous clots that wall off the infected or damaged area; (2) kinins, which reinforce numerous facets of the local inflammatory response; and (3) the complement system, which directly destroys non-self cells by lysing their membranes and furthermore augments other aspects of the inflammatory process. Some of the complement proteins as well as other inflammatory mediators serve as chemotaxins that attract phagocytes to the area. Certain activated complement factors as well as antibodies act as opsonins that link microbes to phagocytic cells, which engulf and destroy them. Phagocytic cells, in addition to destroying microbes intracellularly by phagocytosis, also release chemical mediators that contribute to a wide array of inflammatory activities.

Once the inflammatory response has cleared an area of foreign invaders and tissue debris, tissue repair begins. Perfect repair is possible in regenerative tissues in which cells specific for the tissue undergo cell division to replace lost cells, whereas scar tissue derived from fibroblastic activity replaces lost cells in nonregenerative tissues.

Interferon, a family of related proteins that are nonspecifically released by virus-infected cells, transiently inhibits viral multiplication in other cells to which it binds. It accomplishes its antiviral activity by inducing potential host cells to synthesize enzymes which, upon being activated by viral invasion, can break down viral messenger RNA and inhibit the protein synthesis essential for viral replication. Interferon further enhances the immune activities accomplished by phagocytes, antibodies, natural killer cells, and cytotoxic T cells. These latter two lethal cells attack and destroy both virus-infected cells and cancer cells by lysing their membranes. Natural killer cells nonspecifically attack upon first exposure to an unwanted cell, whereas cytotoxic T cells require prior exposure and preparation before they are capable of assault.

Learning Check (Answers on p. A-28)

A. Multiple choice

1. Which of the following is (are) not part of the nonspecific immune response to foreign invasion?

   a. cytotoxic T cells

   b. interferon

   c. inflammation

   d. natural killer cells

   e. antibody formation

   f. complement system

2. Which of the following is (are) <u>not</u> part of the inflammatory response?

   a. localized vasodilation

   b. migration of neutrophils and macrophages to the site of injury

   c. kinin activation

   d. interferon inhibition of viral replication

   e. formation of interstitial-fluid clots that wall off bacterial invaders

3. Which of the following is (are) <u>not</u> accomplished by a chemical released from activated phagocytes?

   a. decreasing plasma iron so that is is unavailable for bacterial multiplication

   b. stimulating histamine release from mast cells

   c. stimulating the synthesis of viral-blocking enzymes

   d. activating kinins

   e. causing lysis of microbes by punching a hole in their surface membranes

   f. inducing the development of fever

   g. stimulating granulopoiesis

   h. enhancing proliferation and differentiation of B and T lymphocytes

4. Which of the following is (are) <u>not</u> attributable to complement activity?

   a. acting as opsonins

   b. walling-off the inflamed area

   c. serving as chemotaxins

   d. stimulating the release of histamine

   e. forming a membrane attack complex

B. True/False

T/F 1. The early stage of inflammation is predominated by neutrophils.

T/F 2. Phagocytes can destroy foreign microbes only after they have engulfed these invaders.

T/F 3. Endogenous pyrogen (EP), leukocyte endogenous mediator (LEM), and interleukin 1 (IL-1) are all believed to be identical or closely related chemical mediators.

T/F 4. Damaged tissue is always replaced by scar tissue.

T/F 5. Interferon is released only from phagocytic cells that have been invaded by viruses.

T/F 6. Interferon, natural killer cells, and cytotoxic T cells all exert antiviral and anticancer effects.

C. Definitions. Indicate what aspect of the immune defense system is being defined.

1. Sticking of blood-borne leukocytes to the endothelial lining

____

2. Leukocytes squeezing through the capillary pores to exit the vasculature

____

3. Chemical attraction of leukocytes to the site of invasion

____

4. Walled-off structure within which nondestructable offending material is imprisoned

____

5. Collection of phagocytic cells, necrotic tissue, and bacteria

____

6. Chemical that enhances phagocytosis by serving as a link between a microbe and phagocytic cell

____

7. Lymphocytelike cells that nonspecifically destroy virus-infected cells and tumor cells

____

8. Forms a doughnut-shaped complex that imbeds in a microbial surface membrane, causing osmotic lysis of the victim cell

____

9. Localized response to microbial invasion or tissue injury, which is accompanied by swelling, heat, redness, and pain.

____

D. Short answer:

1. Name two categories of drugs that exert anti-inflammatory effects.

   (1) __________________ (2) ______________________

2. Name three means by which microbes can be destroyed in the body without being phagocytized.

   (1) ______________________________________

   (2) ______________________________________

   (3) ______________________________________

3. Distinguish between the two pathways by which the complement system can be activated.

   (1) ____

   (2) ____

Specific Immune Responses

Contents

General Concepts (text page 374)

B Lymphocytes: Antibody-Mediated Immunity (text page 376)

T Lymphocytes: Cell-Mediated Immunity (text page 386)

Immune surveillance against cancer cells involves an interplay among cytotoxic T cells, NK cells, macrophages, and interferon. p. 393

It appears that there is a regulatory loop between the immune system and the nervous and endocrine systems. p. 394

<u>A Closer Look at Exercise Physiology - Exercise: A Help or Hindrance to Immune Defense?</u> p. 396

Section Synopsis

Following initial exposure to a "non-self" entity, such as a microbial invader, specific components of the immune system become especially prepared to selectively attack the particular foreigner. The immune system is able not only to recognize foreign molecules as different from self molecules - so that destructive immune reactions are not unleashed against the body itself - but it can also specifically distinguish between millions of different foreign molecules. The cells of the specific immune system, the lymphocytes, are each uniquely equipped with surface membrane receptors that are able to bind "lock and key" fashion with only one specific complex foreign molecule, which is known as an antigen. The tremendous variation in antigen-detecting ability between different lymphocytes arises from the shuffling around of a few different gene segments, coupled with a high incidence of somatic mutation, during lymphocyte development. Those lymphocytes produced by chance that are able to attack the body's own antigen-bearing cells are eliminated or suppressed so that they are prevented from functioning. In this way, the body is able to "tolerate" (not attack) its own antigens. The major surface antigens of all nucleated cells are known as human leukocyte-association (HLA) antigens, which are coded for by the major histocompatibility complex (MHC), a group of genes with DNA sequences unique for each individual.

There are two broad classes of specific immune responses: antibody-mediated immunity and cell-mediated immunity. In both instances, the ultimate outcome of a particular lymphocyte binding with a specific antigen is destruction of the antigen, but the effector cells, stimuli, and tactics involved are different. Plasma cells derived from B lymphocytes (B cells) are responsible for antibody-mediated immunity, whereas T lymphocytes (T cells) accomplish cell-mediated immunity. B cells develop from a lineage of lymphocytes that originally matured within the bone marrow. The T cell lineage arises from lymphocytes that migrated from the bone marrow to the thymus to complete their maturation. Both B and T cells circulate freely between the blood, lymph, and tissues, but most of the time they reside in self-replacing colonies within the peripheral lymphoid organs, such as the lymph nodes, tonsils, and spleen.

Lymphocytes, either within the lymphoid organs or circulating around on surveillance, react with any foreign target that they specifically "recognize" by attaching to its antigens. Thus activated by antigen, a lymphocyte rapidly proliferates, producing a clone of its own kind that can specifically wage battle against the invader. Thus, the lymphocytes selected for multiplication are the ones that synthesize the antibody that matches the invading antigen. Some of the newly developed lymphocytes do not participate in the attack but become memory cells that lay in waiting, ready to launch a swifter and more forceful attack should the same foreigner ever invade the body again.

B cells and T cells have different targets because of their different requirements for antigen recognition. Each B cell recognizes specific free extracellular antigen that is not associated with cell-bound self-antigens, such as that found on the surface of bacteria. Appropriately, the activated B cell differentiates into a plasma cell, which is specialized to secrete freely circulating antibodies that besiege the freely-existing invading bacteria (or other foreign substance) that induced their production. Antibodies themselves do not directly destroy the foreign material. Instead, they intensify lethal nonspecific immune mechanisms already called into play by the foreign invasion. Antibodies activate the complement system, enhance phagocytosis, and stimulate killer cells.

T cells, in contrast, have a dual binding requirement of foreign antigen in association with self HLA antigens on the surface of one of the body's own cells. Two different types of host cells meet this requirement: (1) virally-invaded host cells; and (2) other immune cells with foreign antigen attached on their surface. As a result of the presence of different classes of self-antigens on the surface of these foreign antigen-bearing host cells, three different types of T cells differentially interact with them. (1) Cytotoxic T cells are able to bind only with virus-infected host cells, whereupon they release toxic substances that kill the infected cell. (2) Helper T cells can only bind with other T cells, B cells, and macrophages that have encountered foreign antigen. Subsequently, helper T cells enhance the immune powers of these other effector cells by secreting specific chemical mediators. (3) Suppressor T cells suppress both T and B cells that have been activated by antigen, thereby preventing the immune system from overresponding and potentially damaging normal host cells. Such differential activation of the various types of lymphocytes assures that the specific immune response that ensues is appropriate to most efficiently dispose of the particular enemy. Moreover, B cells, the various T cells, and macrophages reinforce each other's defense strategies, primarily by releasing a number of important secretory products.

Sometimes cancer cells - once normal cells that somehow through mutation are no longer subject to controls aimed at limiting their growth, multiplication, and position - are able to escape immune surveillance and spread in uncontrolled fashion. Natural killer cells, macrophages, cytotoxic T cells, and the interferon that they collectively secrete normally eradicate newly arisen cancer cells before they have a chance to spread.

Learning Check (Answers on p. A-29)

A. Indicate whether the following characteristics of the specific immune system apply to antibody-mediated immunity, cell-mediated immunity, or both, by writing the appropriate letter in the blank preceding the question below:

a = applies to antibody-mediated immunity

b = applies to cell-mediated immunity

c = applies to both antibody-mediated and cell-mediated immunity

_________ 1. involves secretion of antibodies

_________ 2. mediated by B cells

_________ 3. mediated by T cells

_________ 4. is accomplished by thymic-educated lymphocytes

_________ 5. is triggered by the binding of specific antigens to complementary lymphocyte-bearing receptors

_________ 6. involves formation of memory cells in response to initial exposure to an antigen

_________ 7. is primarily aimed against virally-infected host cells

_________ 8. protects primarily against bacterial invaders

_________ 9. has the capability of directly destroying targeted cells

_________ 10. is involved in rejection of transplanted tissue cells

_________ 11. requires binding of a lymphocyte to a free extracellular antigen

_________ 12. requires dual binding of a lymphocyte with both foreign antigen and self-antigens present on the surface of a host cell

B. Indicate whether the following characteristics apply to the Fab or Fc region of an antibody by writing the appropriate letter in the blank preceding the question using the answer code below:

a = applies to the Fab region

b = applies to the Fc region

________ 1. located in the "arm" regions of an antibody

________ 2. located in the "tail" region of an antibody

________ 3. highly variable between different antibodies of the same class

________ 4. constant between different antibodies of the same class

C. Multiple choice.

1. Which of the following is <u>not</u> accomplished by antibodies?

   a. neutralization of bacterial toxins

   b. direct destruction of foreign cells

   c. activation of the complement system

   d. enhancement of phagocytosis

   e. stimulation of killer (K) cells

2. Which of the following statements regarding T cells is <u>incorrect</u>?

   a. Cytotoxic T cells release chemicals that destroy targeted cells.

   b. Helper T cells enhance the activity of other T cells and B cells.

   c. Suppressor T cells are believed to play an important role in tolerance to self-antigens.

   d. Helper T cells can combine only with host cells bearing both foreign antigen and class I MHC-encoded self-antigens on their surface.

   e. The vast majority of T cells are helper T cells.

3. Which of the following does not play a direct role in immune surveillance against cancer?

   a. B cells

   b. natural killer (NK) cells

   c. macrophages

   d. cytotoxic T cells

   e. interferon

4. Which of the following is not secreted by helper T cells?

   a. B-cell growth factor

   b. T-cell growth factor

   c. interleukin 1

   d. interleukin 2

   e. macrophage migration inhibition factor

5. Which of the following is not a possible cause of autoimmune disease?

   a. reduction in suppressor T cell activity

   b. prolonged treatment with anti-inflammatory drugs

   c. exposure of normally inaccessible self-antigens

   d. modification of normal self-antigens

   e. exposure of the immune system to a foreign antigen almost identical structurally to a self-antigen

D. True/False

T/F 1. Active immunity against a particular disease can be acquired only by actually having the disease.

T/F 2. A secondary response has a more rapid onset, is more potent, and has a longer duration than a primary response.

T/F 3. Type AB blood can be donated to anyone because it lacks both anti-A and anti-B antibodies.

T/F 4. One of the most lethal consequences of mismatched blood transfusions is acute kidney failure caused by blockage of urine-forming structures by hemoglobin precipitation.

T/F 5. The tremendous diversity of antibodies is made possible by genetic recombination coupled with somatic mutation during lymphocyte development.

T/F 6. A single mutation induced by a carcinogen is usually sufficient to convert a normal cell into a cancer cell.

E. Fill-in-the-blank

1. A(n) __________ is a large complex molecule that triggers an immune response against itself.

2. A(n) __________ is a low molecular weight molecule that becomes antigenic by attaching to body proteins.

3. __________ cells derived from activated B lymphocytes are specialized for antibody production.

4. Clumping of foreign cells brought about by the formation of antigen-antibody complexes is known as __________.

5. The __________ theory proposes that a diversity of lymphocytes are produced during development, each preprogrammed to synthesize antibody against only one of an almost limitless variety of antigens.

6. Lymphocytes can only recognize and be activated by antigens that have been processed and presented to them by ________________________.

7. The type of immune cell selectively invaded by AIDS virus is ________________.

8. The __________ is a group of genes that code for unique self-antigens present within the plasma membrane of body cells.

9. __________ is the name applied to these self-antigens.

10. __________ refer collectively to all of the chemical messengers other than antibodies secreted by lymphocytes.

11. __________ immunity is conferred by receipt of preformed antibodies.

12. A mass of transformed cells that is slow-growing, stays put, and does not infiltrate surrounding tissue is known as a __________ tumor, whereas rapidly-growing, invasive masses are called __________ tumors or ______________ . The spreading of mutant cells that have broken away from the parent tumor to other body sites is called ________________.

Immune Diseases **(text page 396)**

Contents

Immune deficiency diseases reduce resistance to foreign invaders. p. 396

Inappropriate immune attacks against harmless environmental substances are responsible for allergies. p. 396

Section Synopsis

There are several ways in which the immune system can go awry. Occasionally, through a deficiency of B or T cells, the immune system fails to defend normally against bacteria or viral infections, respectively. In contrast, in certain instances the immune system becomes overzealous. Instead of limiting its destructive powers to harmful pathogens or cancer cells, the immune system may commit domestic offense by damaging normal body cells. In autoimmune disease, the immune system deliberately turns against one of the person's own tissues that it no longer recognizes and tolerates as self. With immune complex diseases, body tissues are inadvertently destroyed as an overabundance of antigen-antibody complexes activates excessive quantities of lethal complement, which destroys surrounding normal cells as well as the antigen. Allergies occur when the immune system inappropriately launches a symptom-producing, body-damaging attack against an allergen, a normally harmless environmental antigen. In this case, a different type of antibody is brought into play compared with the types of antibodies involved in defense against bacterial invaders. With allergies, IgE is produced, which attaches to mast cells and basophils, causing them to release inflammatory-inducing chemicals such as histamine in the presence of an appropriate allergen. By comparison, with bacterial invasion, IgM and IgG are produced, which attach to and activate complement components and macrophages that kill the bacteria.

Learning Check (Answers on p. A-31)

A. True/False

T/F 1. Histamine is primarily responsible for causing the bronchial constriction associated with asthma.

T/F 2. Eosinophils are attracted to sites involved with delayed allergic reactions.

T/F 3. B lymphocytes are involved with immediate hypersensitivity reactions, whereas T lymphocytes are involved with delayed hypersensitivity reactions.

B. Fill-in-the-blank

1. __________ disease occurs when destructive inflammatory processes "spill over" into normal tissue in the presence of excessive numbers of antigen-antibody complexes.

2. The type of antibodies responsible for inducing allergic manifestations are ____________________.

3. ______________________ refers to the life-threatening allergic phenomenon characterized by severe hypotension and profound bronchial constriction due to the presence of large amounts of chemical mediators in the blood released from mast cells and basophils in response to a particular allergen.

External Defenses (text page 399)

Contents

The skin consists of an outer protective epidermis and an inner connective-tissue dermis. p. 399

Specialized cells in the epidermis produce keratin and melanin and participate in immune defense. p. 402

Protective measures within body cavities that communicate with the external environment discourage pathogen invasion into the body. p. 402

Section Synopsis

The body surfaces exposed to the outside environment - both the outer covering of skin and the linings of internal cavities that communicate with the external environment - serve not only as mechanical barriers to deter would-be pathogenic invaders but play an active role in thwarting entry of bacteria and other unwanted materials as well.

The skin consists of two layers, an outer avascular, keratinized epidermis and an inner connective tissue dermis. The epidermis contains four cell types: melanocytes, keratinocytes, Langerhans cells, and Granstein cells. Melanocytes produce a brown pigment, melanin, the amount of which is responsible for the varying shades of brown skin color. Melanin protects the skin by absorbing harmful ultraviolet radiation. The most abundant cells are the keratinocytes, producers of the tough keratin that forms the outer protective layer of the skin. This physical barrier discourages passage of bacteria and other harmful environmental agents into the body and prevents water and other valuable body substances from escaping. Keratinocytes further serve immunologically by secreting interleukin 1, which enhances post-thymic T cell maturation within the skin. Langerhans cells and Granstein cells also function in specific immunity by respectively presenting antigen to helper T cells and suppressor T cells.

The dermis contains: (1) blood vessels, which nourish the skin and play an important role in regulating body temperature; (2) sensory nerve endings, which provide information about the external environment; and (3) several exocrine glands and hair follicles, which are formed by specialized invaginations of the overlying epithelium. The skin's exocrine glands include sebaceous glands, which produce sebum, an oily substance that softens and waterproofs the skin, and sweat glands, which produce cooling sweat. Hair follicles produce hairs, the distribution and function of which are minimal in humans. Additionally, the skin synthesizes vitamin D in the presence of sunlight.

Besides the skin, the other main routes of entry of potential pathogens into the body are: (1) the digestive system, which is defended by an antimicrobial salivary enzyme, destructive acidic gastric secretions, gut-associated lymphoid tissue, and harmless colonic resident flora; (2) the genitourinary system, which is protected by destructive acidic and particle-entrapping mucus secretions; and (3) the respiratory system, whose defense depends on alveolar macrophage activity and on secretion of a sticky mucus that traps debris, which is subsequently swept out by ciliary action. Other respiratory-defenses include nasal hairs, which filter out large inspired particles; reflex cough and sneeze mechanisms, which expel irritant materials from the trachea and nose, respectively; and the tonsils and adenoids, which defend immunologically.

Learning Check (Answers on p. A-31)

A. Indicate which of the following characteristics apply to the epidermis or dermis of the skin by circling the appropriate letter using the answer code below.

a = applies to the epidermis

b = applies to the dermis

1. is the inner layer of skin a b

2. consists in part of layers of epithelial cells that are dead and flattened a b

3. has no direct blood supply a b

4. contains sensory nerve endings a b

5. contains keratinocytes a b

6. contains melanocytes a b

7. contains rapidly dividing cells a b

8. consists primarily of connective tissue a b

B. True/False

T/F 1. Severe burns of the skin can result in life-threatening circulatory disturbances.

T/F 2. Adipose tissue is located within the hypodermis.

T/F 3. Saliva is destructive to bacteria because it is highly acidic.

T/F 4. The large intestine's normal microbial population helps defend against infection within the lower intestine.

T/F 5. Debris trapped on the sticky mucus lining the respiratory airways is primarily cleared away by the aveolar macrophages.

T/F 6. A sneeze expels irritant material from the trachea.

C. Matching

_______	1.	secrete a pigment responsible for varying shades of brown color in the skin	a. keratinocytes
_______	2.	migrate to the skin from the bone marrow	b. melanocytes
_______	3.	secrete interleukin 1	c. Langerhans cells
_______	4.	produce the tough protective surface of the skin	d. Granstein cells
_______	5.	secrete a substance that absorbs harmful ultraviolet rays	
_______	6.	present antigen to helper T cells	
_______	7.	present antigen to suppressor T cells	
_______	8.	the most abundant cell type in the skin	
_______	9.	produce hair and nails	

D. Short answer

1. List two functions of dermal blood vessels.

   a.
   b.

2. Indicate the secretory product and function of the following three specialized skin structures.

Name	Secretory Product	Function
a. sweat glands		
b. sebaceous glands		
c. hair follicles		

Chapter in Perspective **(text page 404)**

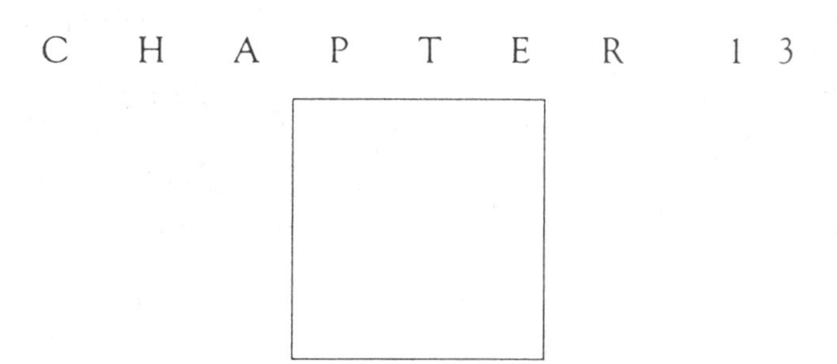

Respiratory System

Introduction (text page 406)

Contents

Section Synopsis

Internal respiration refers to the intracellular metabolic reactions that utilize O_2 and produce CO_2 during energy-yielding oxidation of nutrient molecules. The ratio of CO_2 produced to O_2 consumed is known as the respiratory quotient, which averages 0.8 on a typical American diet. External respiration encompasses the various steps involved in the transfer of O_2 and CO_2 between the external environment and tissue cells. The respiratory and circulatory systems function together to accomplish external respiration.

The respiratory system accomplishes exchange of air between the atmosphere and the lungs by means of the process of ventilation or breathing. Exchange of O_2 and CO_2 between the air in the lungs and the blood in the pulmonary capillaries takes place across the extremely thin walls of the air sacs, or alveoli, which collectively provide a tremendous surface area available for transfer of these gases. The respiratory airways that conduct air from the atmosphere to the gas-exchanging portion of the lungs include the nasal passages, pharynx, larynx, trachea, and a highly branching system of bronchi that lead into ever smaller and more numerous bronchioles, which finally terminate at the alveoli. The lungs consist of these branching conducting airways (beyond the two main bronchi, one entering each lung), the alveoli and their surrounding pulmonary capillaries, and an abundance of elastic connective tissue. The lungs are housed within the closed compartment of the thorax, the volume of which can be changed by contractile activity of surrounding respiratory muscles. A double-walled closed sac lined with lubricating fluid, the pleural sac, separates each lung from the surrounding thoracic wall.

Learning Check (Answers on p. A-33)

A. Fill-in-the-blank

1. The ________________ serve as the conducting portion of the respiratory system and the ____________________ are the gas-exchanging portion.

2. The respiratory quotient is the ratio of _____________ to _______________________.

3. The exchange of O_2 and CO_2 between the external environment and tissue cells is known as____________ .

4. The ________________ serves as a common passageway for both the respiratory and digestive systems.

5. The alveolar wall is (how many) ________ cell layer(s) thick. The wall of the pulmonary capillaries surrounding an alveolus is (how many) cell ________ layer(s) thick.

6. __________________ refers to air flow between adjacent alveoli through Pores of Kohn.

7. The most profound changes in thoracic volume can be accomplished by contraction of the ______________ .

8. The ___________________ is a double-walled closed sac that separates each lung from the thoracic wall and other surrounding structures.

B. True/False

T/F 1. All steps of external respiration are accomplished by the respiratory system.

T/F 2. Alternate contraction and relaxation of muscles within the lung tissue accomplishes the act of breathing.

T/F 3. The respiratory airways filter, warm, and humidify incoming air.

T/F 4. The sole purpose of the vocal cords is related to their role in initiating the sounds of speech.

T/F 5. All respiratory airways are held open by cartilaginous rings.

T/F 6. Air flow through the smaller bronchioles can be adjusted by varying the contractile activity of the smooth muscle within their walls.

Respiratory Mechanics (text page 412)

Contents

Section Synopsis

Ventilation is the process of cyclically moving air in and out of the lungs so that old alveolar air that has already participated in exchange of O_2 and CO_2 with the pulmonary capillary blood can be switched for fresh atmospheric air. Ventilation is mechanically accomplished by alternately shifting the direction of the pressure gradient for air flow between the atmosphere and alveoli through cyclical expansion and recoil of the lungs. Alternate contraction and relaxation of the respiratory muscles indirectly produces periodic inflation and deflation of the lungs by cyclically expanding and compressing the thoracic cavity. The lungs follow the movements of the thoracic cavity because of: (1) the surface-tension forces of the intrapleural fluid separating the lungs from the chest wall; (2) the lung-stretching transmural pressure gradient across the lung wall; and (3) the elastic-recoil properties of the lungs, which cause them to passively deflate once stretching forces are removed.

During inspiration, as contraction of the diaphragm and other inspiratory muscles enlarges the thorax, the lungs are likewise forced to expand. The intraalveolar pressure falls below atmospheric pressure as the air within the lungs is rarefied upon lung expansion. The pressure gradient thus created between the atmosphere and lungs causes air to enter the lungs. During expiration, the inspiratory muscles relax, allowing the thorax and lungs to recoil to their preinspiratory size. As the air within the lungs is compressed during lung recoil, the intraalveolar pressure rises above atmospheric pressure, establishing a pressure gradient for the outward flow of air. Because energy is required for contraction of the inspiratory muscles, inspiration is an active process, but expiration is passive during quiet breathing because it is accomplished by relaxation of inspiratory muscles at no energy expense. For more forceful active expiration, contraction of the expiratory muscles, namely the abdominal muscles, further decreases the dimensions of the thorax and lungs, which further increases the intraalveolar-to-atmospheric pressure gradient. The larger the gradient between the alveoli and atmosphere in either direction, the larger the air flow rate, because air continues to flow until the intraalveolar pressure equilibrates with atmospheric pressure.

Besides being directly proportional to the pressure gradient, air flow rate is also inversely proportional to airway resistance. Because airway resistance, which depends on the caliber of the conducting airways, is normally very low, air flow rate usually depends primarily on the pressure gradient established between the alveoli and atmosphere. If airway resistance is pathologically increased by chronic obstructive pulmonary disease, the pressure gradient must be correspondingly increased by more vigorous respiratory muscle activity to maintain a normal air flow rate.

The lungs are able to be stretched to varying degrees during inspiration and then to recoil to their preinspiratory size during expiration because of their elastic behavior. Pulmonary compliance refers to the distensibility of the lungs - how much they stretch in response to a given change in the transmural pressure gradient. Elastic recoil refers to the phenomenon of the lungs snapping back to their resting position during expiration. Pulmonary elastic behavior is dependent on the elastic connective tissue meshwork within the lungs and on alveolar surface tension/pulmonary surfactant interaction. Alveolar surface tension, which is due to the attractive forces between the surface water molecules in the liquid film lining each alveolus, tends to resist the alveolus being stretched upon inflation (decreases compliance) and tends to return it back to a smaller surface area during deflation (increases elastic recoil). If the alveoli were lined by H_2O alone, the surface tension would be so great that the lungs would be poorly compliant and would tend to collapse. Fortunately, Type II alveolar cells secrete pulmonary surfactant, a phospholipoprotein that intersperses between the water molecules. Pulmonary surfactant lowers the alveolar surface tension, thereby increasing the compliance of the lungs and counteracting the tendency for small alveoli to collapse. Lung stability is also promoted by alveolar interdependence, which refers to the mutual tethering effect provided by neighboring alveoli. The energy required to accomplish ventilation is necessary to expand the thorax and indirectly the lungs against their elastic recoil forces plus to overcome airway resistance.

Normally the lungs operate at "half-full." The lung volume typically varies between about 2 to 2 1/2 liters as an average tidal volume of 500 ml of air is moved in and out with each breath. The lungs can be filled an additional 3 liters upon maximal inspiratory effort or emptied to about 1 liter upon maximal expiratory effort. The lungs cannot be completely emptied because of the transmural pressure gradient and because the small airways collapse during forced expirations at low lung volumes, precluding further outflow of air.

The amount of air moved in and out of the lungs in one minute, the pulmonary ventilation, is equal to tidal volume x respiratory rate. However, not all of the air moved in and out is available for O_2 and CO_2 exchange with the blood because part of it occupies the conducting airways, known as the anatomic dead space, rather than exchanging between the atmosphere and alveoli. Alveolar ventilation, the volume of air exchanged between the atmosphere and alveoli in one minute, is a measure of the air actually available for gas exchange with the blood. Alveolar ventilation equals (tidal volume minus the dead space volume) x respiratory rate. Alveolar ventilation can be markedly varied by changes in breathing pattern (i.e., changing the depth and rate of breathing), even though there is little change in total pulmonary ventilation.

If local changes in ventilation or perfusion arise, the subsequent changes in local chemical factors bring about adjustments in airway and vessel caliber by acting directly on bronchiolar and arteriolar smooth muscle to rematch air flow and blood flow as much as possible.

Learning Check (Answers on p. 34)

A. Fill-in-the-blank.

1. The three forces that tend to keep the alveoli open are ____________________, ________________________, and _________________________.

2. The two forces that promote alveolar collapse are _________________________ and _________________________.

3. ___________________________ is a respiratory disease characterized by collapse of the smaller airways and a breakdown of alveolar walls.

4. ___________________________ is a measure of the magnitude of change in lung volume accomplished by a given change in transmural pressure gradient.

5. ___________________________ refers to the phenomenon of the lungs snapping back to their resting size after having been stretched.

6. The maximum volume of air that can be moved in and out of the lungs in a single breath is known as ___________________________.

7. The volume occupied by the conducting airways is known as ___________________________.

B. True/False

T/F 1. The alveoli normally empty completely during maximal expiratory efforts.

T/F 2. An increase in airway resistance or an increase in alveolar surface tension both increase the work of breathing.

T/F 3. Smaller alveoli display a larger inward-directed pressure than do larger alveoli.

T/F 4. Alveolar ventilation always increases when pulmonary ventilation increases.

T/F 5. The 500 ml of air that is inspired is the same 500 ml of air that enters the alveoli during a single breath.

C. Indicate the effect the following factors in question have in regulating air flow and blood flow by circling the appropriate word.

1. parasympathetic stimulation	dilates	constricts	bronchioles
2. epinephrine	dilates	constricts	bronchioles
3. local increase in alveolar CO_2	dilates	constricts	bronchiole supplying this alveolus
4. local increase in alveolar O_2	dilates	constricts	pulmonary arteriole supplying this alveolus

D. Indicate the relationship that exists between the two items in question by circling:

> (greater than), < (less than), or = (equal to)

1. The size of the thoracic cavity when the diaphragm is contracting is (>, <, =) the size of the thoracic cavity when the diaphragm is relaxed.

2. Lung volume before the diaphragm contracts is (>,<,=) lung volume after the diaphragm contracts.

3. Intra-alveolar pressure during inspiration when air is flowing into the lungs is (>,<,=) intra-alveolar pressure before the onset of inspiration when no air is flowing.

4. Intra-alveolar pressure at the end of inspiration is (>,<,=) atmospheric pressure.

5. The number of molecules of air in the lungs at the onset of inspiration is (>,<,=) the number of molecules of air in the lungs at the end of inspiration.

6. Intra-alveolar pressure during expiration when air is flowing out of the lungs is (>,<,=) intra-alveolar pressure before the onset of expiration when no air is flowing.

7. Intra-alveolar pressure at the end of expiration, prior to the onset of inspiration, is (>,<,=) atmospheric pressure.

8. Intrapleural pressure at any point during the respiratory cycle is (>,<,=) intra-alveolar pressure.

9. The number of molecules of air in the lungs at the onset of expiration is (>,<,=) the number of molecules of air in the lungs at the end of expiration.

10. Intra-alveolar pressure upon relaxation of the diaphragm is (>,<,=) intra-alveolar pressure upon relaxation of the diaphragm plus contraction of the abdominal muscles.

11. The size of the thoracic cavity during contraction of the internal intercostal muscles is (>,<,=) the size of the thoracic cavity during contraction of the external intercostal muscles.

12. During quiet breathing, energy expenditure during inspiration is (>,<,=) energy expenditure during expiration.

13. Air flow during passive expiration is (>,<,=) air flow during active expiration.

14. Intrapleural pressure in the presence of pneumothorax is (>,<,=) atmospheric pressure.

15. The pressure gradient needed to move a normal tidal volume in a healthy individual is (>,<,=) the pressure gradient needed to move a normal tidal volume in a person with obstructive lung disease.

16. The total lung capacity of a person with restrictive lung disease is (>,<,=) the total lung capacity of a person with obstructive lung disease.

17. The residual volume of a person with restrictive lung disease is (>,<,=) the residual volume of a person with obstructive lung disease.

18. The FEV_1/VC % of a person with restrictive lung disease is (>,<,=) the FEV_1/VC % of a person with obstructive lung disease.

19. Alveolar surface tension of normal lungs is (>,<,=) alveolar surface tension of an infant with newborn respiratory distress syndrome.

Gas Exchange **(text page 434)**

Contents

Section Synopsis

Oxygen and carbon dioxide move across body membranes by passive diffusion down partial pressure gradients. No active transport mechanisms are necessary for gaseous exchange. The partial pressure exerted by a particular gas in the air is directly proportional to the percentage of that gas in the total air mixture. The partial pressure exerted by a gas in the blood is directly proportional to the concentration of that gas dissolved in thc blood.

Net diffusion of O_2 occurs first between the alveoli and the blood and then between the blood and the tissues as a result of the O_2 partial pressure gradients created by continuous utilization of O_2 in the cells and continuous replenishment of fresh alveolar O_2 provided by pulmonary ventilation. Net diffusion of CO_2 occurs in the reverse direction, first between the tissues and the blood and then between the blood and the alveoli, as a result of the CO_2 partial pressure gradients created by continuous production of CO_2 in the cells and continuous removal of alveolar CO_2 through the process of pulmonary ventilation.

The average alveolar partial pressures remain fairly constant throughout the respiratory cycle at a Po_2 of 100 mm Hg and a Pco_2 of 40 mm Hg. The Po_2 of old systemic venous blood entering the lungs via the pulmonary capillaries averages 40 mm Hg and the Pco_2 46 mm Hg. Thus, the partial pressure gradients favor the movement of O_2 from the alveolar air into the pulmonary capillary blood and of CO_2 in the opposite direction. The alveolar-pulmonary capillary barrier separating the air and blood is ideally suited to facilitate gas exchange because of its tremendous surface area and thinness. Normally the blood partial pressures have equilibrated with alveolar partial pressures before the blood leaves the pulmonary capillay bed, so the blood that returns from the lungs to the heart and is subsequently pumped out as systemic arterial blood has a Po_2 of 100 mm Hg and Pco_2 of 40 mm Hg, the same as alveolar air. However, pulmonary diseases that impede gas exchange can prevent the blood from completely "filling up" on O_2 and "dumping off" CO_2 as it passes through the lungs.

The average cellular Po_2 is 40 mm Hg and Pco_2 is 46 mm Hg, establishing partial pressure gradients favoring the movement of O_2 from the blood into the cells and the movement of CO_2 from the cells into the blood. This process continues until the venous blood leaving the tissues has equilibrated with the cells at a Po_2 of 40 mm Hg and a Pco_2 of 46 mm Hg. This blood is returned to the heart and pumped to the lungs for exchange with alveolar air once again.

If the cells are more rapidly metabolizing, their Po_2 falls and their Pco_2 increases correspondingly, the result of which is larger partial pressure gradients that promote a greater net diffusion of O_2 from the atmosphere to the cells and of CO_2 from the cells to the atmosphere. Therefore, gas transfer normally automatically matches metabolic requirements by simply passive means.

Learning Check (Answers of p. A-35)

A. True/False

T/F 1. Alveolar partial pressures do not fluctuate to any extent between inspiration and expiration.

T/F 2. The surface area available for exchange within the lungs actually increases during exercise.

T/F 3. Less time is available for gas exchange when cardiac output increases.

T/F 4. The diffusion coefficients for O_2 and CO_2 are equal.

B. Indicate the effect the following changes would have on the rate of gas transfer by circling the appropriate letter using the answer code below:

a = this change would increase the rate of gas transfer

b = this change would decrease the rate of gas transfer

c = this change would have no effect on the rate of gas transfer

a b c 1. the effect of pulmonary fibrosis on O_2 and CO_2 exchange within the lungs

a b c 2. the effect of emphysema on O_2 and CO_2 exchange within the lungs

a b c 3. the effect of a fall in atmospheric Po_2 on O_2 exchange within the lungs

a b c 4. the effect on O_2 exchange in the lungs of replacing part of the nitrogen with helium so that the inspired air consists of 60% N, 20% He, and 20% O_2

a b c 5. the effect of increased metabolism of a cell on O_2 and CO_2 exchange between the cell and blood

a b c 6. the effect of tissue edema on O_2 and CO_2 exchange between the surrounding cells and blood

a b c 7. the effect of reduced systemic venous Po_2 on O_2 exchange within the lungs

C. If a person ascended a mountain where the atmospheric pressure was only 500 mm Hg, what would the Po_2 of the air be, assuming that the air consisted of 20% O_2?______________

D. Indicate the O_2 and CO_2 partial pressure relationships that are important in gas exchange by circling > (greater than), < (less than), or = (equal to) as appropriate in each of the following statements.

1. Po_2 in blood entering the pulmonary capillaries is (>, <, or =) Po_2 in the alveoli.

2. Pco_2 in blood entering the pulmonary capillaries is (>, <, or =) Pco_2 in the alveoli.

3. Po_2 in the alveoli is (>, <, or =) Po_2 in blood leaving the pulmonary capillaries.

4. Pco_2 in the alveoli is (>, <, or =) Pco_2 in blood leaving the pulmonary capillaries.

5. Po_2 in blood leaving the pulmonary capillaries is (>, <, or =) Po_2 in blood entering the systemic capillaries.

6. Pco_2 in blood leaving the pulmonary capillaries is (>, <, or =) Pco_2 in blood entering the systemic capillaries.

7. Po_2 in blood entering the systemic capillaries is (>, <, or =) Po_2 in the tissue cells.

8. Pco_2 in blood entering the systemic capillaries is (>, <, or =) Pco_2 in the tissue cells.

9. Po_2 in the tissue cells is (>, <, or approximately =) Po_2 in blood leaving the systemic capillaries.

10. Pco_2 in the tissue cells is (>, <, or approximately =) Pco_2 in the blood leaving the systemic capillaries.

11. Po_2 in blood leaving the systemic capillaries is (>, <, or =) Po_2 in blood entering the pulmonary capillaries.

12. Pco_2 in blood leaving the systemic capillaries is (>, <, or =) Pco_2 in blood entering the pulmonary capillaries.

Gas Transport (text page 440)

Contents

Section Synopsis

Oxygen and carbon dioxide must be transported between their sites of exchange at the pulmonary capillary and systemic capillary levels by means of the blood. Because these gases are not very soluble in the blood, they must be transported primarily by mechanisms other than simply being physically dissolved. In the case of O_2 transport in the blood, only 1.5% of the O_2 is physically dissolved and 98.5% is chemically bound to hemoglobin (Hb). Hemoglobin and oxygen reversibly form oxyhemoglobin (HbO_2). The loading reaction, $Hb + O_2 \rightarrow HbO_2$, occurs in the pulmonary capillaries as O_2 is picked up. The unloading reaction, $HbO_2 \rightarrow Hb + O_2$, occurs in the systemic capillaries as O_2 is released to the tissue cells.

The primary factor that determines the extent to which Hb and O_2 are combined, or the % Hb saturation, is the Po_2 of the blood. The relationship between blood Po_2 and % Hb saturation is such that in the Po_2 range within the pulmonary capillaries, Hb is still almost fully saturated even if the blood Po_2 falls as much as 40%. This provides a margin of safety in assuring near normal O_2 delivery to the tissues despite a substantial reduction in arterial Po_2. On the other hand, in the Po_2 range within the systemic capillaries, large increases in Hb unloading occur in response to a small local decline in blood Po_2 associated with increased cellular metabolism, thereby providing more O_2 to match the increased tissue needs. The exact % Hb saturation at any given Po_2 is influenced by the affinity or bond strength between Hb and O_2. The affinity of Hb for O_2 is reduced and thus O_2 unloading enhanced in response to the increased CO_2, acidity, and temperature found in the tissue environment. Similarly, increased 2,3 diphosphoglycerate (DPG) production by red blood cells in response to reduced O_2 availability allows a greater percentage of O_2 to be liberated at the tissues by decreasing Hb affinity for O_2. The poisonous gas carbon monoxide (CO), on the other hand, seriously impairs O_2 loading and unloading by: (1) tenaciously tying up a disproportionate amount of Hb so that less Hb is available for O_2 transport; and (2) interfering with the release of O_2 from the Hb that is carrying O_2.

In addition to transporting O_2, Hb, by acting as a storage depot, facilitates the net transfer of large amounts of O_2 between the blood and surrounding tissues. In the lungs, Hb keeps the blood Po_2 low by removing O_2 from solution. Since O_2 bound to Hb does not contribute to the blood Po_2, an air-to-blood Po_2 gradient is maintained despite the transfer of a large quantity of O_2. This gradient favoring O_2 pick-up is maintained until Hb is loaded with O_2 to the greatest extent possible. At the tissue level, the unloading of O_2 from Hb keeps the blood Po_2 high, maintaining a blood-to-tissue gradient that favors the transfer of large amounts of O_2 into the tissue cells for their consumption.

Carbon dioxide that is picked up at the systemic capillaries is transported in the blood by three methods: (1) 10% physically dissolved; (2) 30% bound to Hb; and (3) 60% as bicarbonate (HCO_3^-). The erythrocytic enzyme, carbonic anhydrase, catalyzes the conversion of CO_2 to HCO_3^- according to the reaction:

$CO_2 + H_2O \overset{c.a.}{\rightleftharpoons} H_2CO_3 \rightleftharpoons H^+ + HCO_3^-$. The generated H^+ binds to Hb. Unloading of O_2 from Hb at the tissue level increases the affinity of Hb for both CO_2 and CO_2-generated H^+. Bicarbonate diffuses out of the red blood cells into the plasma down its concentration gradient. Chloride shifts into the red cells to maintain electric neutrality. These reactions are all reversed in the lungs as CO_2 is eliminated to the alveoli.

Learning Check (Answers on p. A-35)

A. Listing

1. List the methods of O_2 and CO_2 transport, indicating the % of gas carried by each method.

a. Methods of O_2 transport

________________ ____%

________________ ____%

b. Methods of CO_2 Transport

________________ ____%

________________ ____%

________________ ____%

2. List the substances that can combine with Hb.

__________, __________, __________, __________

B. Fill-in-the-blank.

1. The primary factor that determines the % Hb saturation is the ______________.

2. ______________ is the erythrocytic enzyme responsible for catalyzing the conversion of CO_2 into HCO_3^-.

3. ______________ shifts into the red blood cells to maintain electric neutrality when HCO_3^- moves out of the cells down its concentration gradient.

4. The ____________ effect refers to the reduced affinity of Hb for O_2 in the presence of increased CO_2 and H^+.

5. The ________________ effect refers to the increased affinity of Hb for CO_2 and H^+ after O_2 unloading.

6. ______________________ refers to a ventilation rate that exceeds the metabolic needs of the body.

C. True/False

T/F 1. In the plateau region of the Hb-O_2 curve, a large decrease in Po_2 results in a small decrease in Hb saturation, whereas in the steep portion of the curve a small decrease in Po_2 results in a large decrease in % Hb saturation.

T/F 2. Hemoglobin, by acting as a storage depot, plays an important role in permitting the transfer of large quantities of O_2 between the blood and surrounding tissues down Po_2 gradients because the O_2 bound to Hb does not directly contribute to the blood Po_2.

T/F 3. The combination of Hb and CO_2 is known as carboxyhemoglobin.

T/F 4. Hemoglobin has a higher affinity for O_2 than for any other substance.

T/F 5. Hypercapnia always accompanies hypoxia.

D. Multiple Choice

1. Which of the following factors shift(s) the Hb-O_2 curve to the right, thereby enhancing O_2 unloading?

   a. increased CO_2
   b. increased CO
   c. increased H^+
   d. increased temperature
   e. increased DPG

2. Which of the following reactions take(s) place at the pulmonary capillaries?

   a. $Hb + O_2 \rightarrow HbO_2$
   b. $CO_2 + H_2O \rightarrow H_2CO_3 \rightarrow H^+ + HCO_3^-$
   c. $Hb + CO_2 \rightarrow HbCO_2$
   d. $HbH \rightarrow Hb + H^+$

E. Matching (more than one answer may apply)

_____ 1. due to reduced O_2-carrying capacity of the blood

_____ 2. present in severe emphysema

_____ 3. cannot occur when breathing atmospheric air at sea level

_____ 4. cells are unable to utilize O_2 available to them

_____ 5. low arterial Po_2 due to reduced O_2 uptake at the pulmonary level

_____ 6. blood Po_2 is increased but % Hb saturation is normal

_____ 7. may accompany congestive heart failure

_____ 8. causes respiratory alkalosis

_____ 9. present in CO poisoning

_____10. present in cyanide poisoning

_____11. too little oxygenated blood is delivered to the tissues

_____12. change in arterial Pco_2 accompanying hypoventilation

_____13. causes respiratory acidosis

_____14. change in arterial Pco_2 accompanying hyperventilation

a. hypoxic hypoxia

b. anemic hypoxia

c. circulatory hypoxia

d. histotoxic hypoxia

e. hyperoxia

f. hypercapnia

g. hypocapnia

Control of Respiration (text page 449)

Contents

Section Synopsis

The two distinct aspects of ventilation subject to neural control are: (1) rhythmic cycling between inspiration and expiration; and (2) regulation of the magnitude of ventilation, which in turn depends on control of respiratory rate and depth of tidal volume. Respiratory rhythm is primarily established by pacemaker activity displayed by inspiratory neurons located in the respiratory control center in the medulla of the brain stem. When these inspiratory neurons autonomously fire, impulses ultimately reach the inspiratory muscles to bring about inspiration. When the inspiratory neurons cease firing, the inspiratory muscles relax and expiration takes place. If active expiration is to occur, the expiratory muscles are activated via output from the medullary expiratory neurons at this time. This basic rhythm is smoothed out by a balance of activity in the apneustic and pneumotaxic centers located higher in the brain stem in the pons. The apneustic center prolongs inspiration, whereas the more powerful pneumotaxic center limits inspiration. Another factor that helps keep the duration of inspiration and expiration in balance is negative feedback from the expiratory neurons, which are activated by the inspiratory neurons themselves as a self-limiting mechanism. At tidal volumes greater than one liter, overinflation of the lungs is prevented by the Hering-Breuer reflex, which limits further inspiration by means of activating pulmonary stretch receptors that reflexly inhibit the inspiratory neurons.

Three chemical factors play a role in determining the magnitude of ventilation: the Pco_2, Po_2 and H^+ concentration of the arterial blood. The dominant factor in the minute-to-minute regulation of ventilation is the arterial Pco_2. An increase in arterial Pco_2 is the most potent chemical stimulus for increasing ventilation. Changes in arterial Pco_2 alter ventilation primarily by bringing about corresponding changes in the brain ECF H^+ concentration to which the central chemoreceptors are exquisitely sensitive. Of lesser importance is reflex stimulation of the respiratory center via a weak response of the peripheral chemoreceptors (the carotid and aortic bodies) to an elevated arterial Pco_2. The peripheral chemoreceptors are more responsive to an increase in arterial H^+ concentration, which likewise reflexly brings about increased ventilation. Conversely, a fall in arterial H^+ concentration reflexly suppresses ventilation via the peripheral chemoreceptors. The resultant adjustments in arterial H^+-generating-CO_2 are important in maintaining the acid-base balance of the body. The peripheral chemoreceptors also respond to a marked reduction in arterial Po_2 (<60 mm Hg) to reflexly stimulate the respiratory center. This serves as an emergency mechanism to increase respiration when the arterial Po_2 levels fall below the safety range provided by the plateau portion of the Hb-O_2 curve. This is especially important since the central chemoreceptors and respiratory center are directly depressed by a dangerously low Po_2.

Exercise is a profound respiratory stimulant, although the underlying mechanisms for the abrupt and pronounced increase in ventilation that meets the need for increase O_2 uptake and CO_2 removal are not fully understood.

Several factors modify ventilation to serve purposes other than O_2 uptake and CO_2 removal. These include coughing, sneezing, and other protective reflexes; pain; emotional expression such as laughing and crying; and voluntary control.

Apnea is a transient cessation of breathing, occurring most frequently during sleep in individuals, especially infants, in whom the respiratory control mechanisms are less responsive than normal. Dyspnea is an uncomfortable subjective sensation that ventilation is inadequate.

Learning Check (Answers on p. A-36)

A. True/False

T/F 1. Rhythmicity of breathing is brought about by pacemaker activity displayed by the respiratory muscles.

T/F 2. The expiratory neurons send impulses to the motor neurons controlling the expiratory muscles during normal quiet breathing.

T/F 3. The peripheral chemoreceptors are not activated during carbon monoxide poisoning despite the fact that the total O_2 content in the blood can become lethally low.

T/F 4. Administering O_2 to patients with severe chronic lung disease may markedly depress their drive to breathe.

T/F 5. Arterial Po_2 remains normal or may even increase slightly during exercise despite the fact that O_2 consumption by the tissues is greatly increased.

T/F 6. Respiration is reflexly inhibited by alkalosis caused by a reduction in concentration of non-carbon-dioxide generated H^+, the result of which is accumulation of H^+-generating carbon dioxide to restore the acid-base balance toward normal.

T/F 7. Voluntary control of breathing is accomplished by the cerebral cortex sending impulses to the respiratory center to modify its output.

B. Fill-in-the-blank

1. The primary respiratory control center that provides output to the respiratory muscle is located in the_______________.

2. The apneustic and pneumotaxic centers are located in the _________________.

3. The_______________ nerve supplies the diaphragm.

4. The DRG consists mostly of _______________________ neurons.

5. The expiratory neurons are located in the _________________ .

6. ______________________ is normally the most important input in regulating the magnitude of ventilation under resting conditions.

7. The peripheral chemoreceptors include the ________________ and ___________ bodies.

8. ______________ is the transient cessation of breathing.

9. The subjective sensation of not getting enough air is known as _______________ .

C. Multiple choice

1. Which of the following does <u>not</u> contribute to the cessation of inspiration?

   a. The spontaneous discharge of the inspiratory neurons starts to decline on its own due to the cyclical nature of inspiratory neuron pacemaker activity.

   b. The apneustic center inhibits the inspiratory neurons.

   c. The expiratory neurons inhibit the inspiratory neurons via a self-limiting negative feedback loop initiated by the inspiratory neurons.

   d. Pulmonary stretch receptors activated at large lung volumes reflexly inhibit the inspiratory neurons.

   e. The pneumotaxic center limits inspiration.

2. Which of the following does <u>not</u> contribute to the profound and abrupt increase in ventilation during exercise?

   a. Movement of the limbs

   b. Epinephrine release

   c. Heat production by exercising muscles

   d. Increased production of CO_2

   e. Stimulation of the respiratory center by motor areas of the cerebral cortex.

D. Circle the correct answer using the following answer code:

a = peripheral chemoreceptors

b = central chemoreceptors

c = both the peripheral and central chemoreceptors

d = neither the peripheral or central chemoreceptors

1. stimulated by an arterial Po_2 of 80 mm Hg a b c d

2. stimulated by an arterial Po_2 of 55 mm Hg a b c d

3. directly depressed by an arterial Po_2 of 55 mm Hg a b c d

4. weakly stimulated by an elevated arterial Pco_2 a b c d

5. strongly stimulated by an elevated brain ECF H^+ induced by an elevated arterial Pco_2 a b c d

6. stimulated by an elevated arterial H^+ concentration a b c d

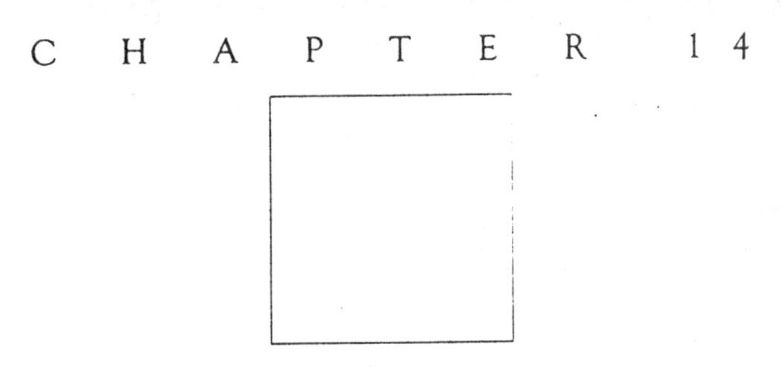

URINARY SYSTEM

Introduction (text page 462)

Contents

The kidneys perform a variety of functions aimed at maintaining homeostasis. p. 463

The kidneys form the urine; the remainder of the urinary system is ductwork to carry the urine to the outside. p. 464

The nephron is the functional unit of the kidney. p. 465

The three basic renal processes are glomerular filtration, tubular reabsorption, and tubular secretion. p. 467

Section Synopsis

The cells of the body can survive and function only if they are bathed by an internal fluid environment of constant composition and volume. The kidneys, which are primarily responsible for maintaining the stability of this internal environment, act on the plasma to eliminate unwanted plasma constituents in the urine while conserving materials of value to the body. The plasma, in turn, is in equilibrium with the interstitial fluid, the true internal fluid environment that bathes the cells. A pair of kidneys form the urine, and the remainder of the urinary system serves as plumbing to eliminate the urine from the body. After urine is produced, it is collected within the renal pelvis of each kidney, then transported through the ureters to the urinary bladder, which temporarily stores the urine until it can be emptied to the exterior through the urethra.

The urine-forming functional unit of the kidneys is the nephron, which is composed of interrelated vascular and tubular components. The vascular component consists of two capillary networks in series, the first being the glomerulus, a ball-like tuft of capillaries that filters large volumes of protein-free plasma into the tubular component. The second capillary network consists of the peritubular capillaries, which wind around the tubular component. The peritubular capillaries nourish the renal tissue and participate in exchanges between the tubular fluid and plasma. The tubular component begins with Bowman's capsule, which cups around the glomerulus to catch the filtrate, then continues a specific tortuous course to ultimately empty into the renal pelvis. As the filtrate passes through the various regions of the tubule, it is modified by the cells lining the tubules to return to the plasma only those materials necessary for maintaining the proper ECF composition and volume. What is left behind in the tubules is excreted as urine.

The kidneys perform three basic processes in carrying out their regulatory and excretory functions: (1) glomerular filtration, the nondiscriminating movement of protein-free plasma from the blood into the tubules; (2) tubular reabsorption, the selective transfer of specific constituents in the filtrate back into the blood of the peritubular capillaries; and (3) tubular secretion, the highly specific movement of selected substances from the peritubular capillary blood into the tubular fluid. There are two means by which materials can enter the renal tubules - glomerular filtration and tubular secretion - and two routes by which materials can leave the tubules - tubular reabsorption back into the blood or elimination from the body via urine excretion. Everything that is filtered or secreted but not reabsorbed is excreted as urine. In general, filtered substances of value to the body are reabsorbed and conserved, whereas unwanted portions of the tubular fluid, such as metabolic wastes, remain in the tubule to be excreted in the urine. A few unwanted substances may even be added to the tubular filtrate by the process of tubular secretion to facilitate their removal from the body. Urine composition and volume vary markedly in the process of maintaining constancy in ECF composition and volume.

Learning Check (Answers on p. A-38)

A. True/False

T/F 1. The kidneys are the organs that are primarily responsible for maintaining constancy of the volume and electrolyte composition of the internal fluid environment.

T/F 2. The glomerular filtrate is almost identical in composition to the plasma.

T/F 3. The kidneys keep the urine volume and composition essentially constant.

T/F 4. The vast majority of the filtered fluid is reabsorbed.

T/F 5. In general, the substances in the filtrate that need to be conserved are selectively reabsorbed whereas the unwanted substances that need to be eliminated fail to be reabsorbed.

B. Multiple choice

1. Which of the following is <u>not</u> accomplished by the kidneys? The kidneys :

   a. contribute significantly to long-term regulation of arterial blood pressure by maintaining the proper plasma volume

   b. act directly on the interstitial fluid, the fluid that bathes the cells, to maintain constancy in its composition

   c. excrete the metabolic waste products

   d. assist in maintaining the proper acid-base balance of the body

   e. secrete several hormones

2. Which of the following is <u>not</u> associated with juxtamedullary nephrons?

a. glomeruli located in medulla

b. long loops of Henle

c. peritubular capillaries form vasa recta

d. collecting duct in medulla

e. play important role in the ability of the kidneys to produce urine of varying concentration

C. Fill-in-the-blank

1. The functional unit of the kidneys is the ____________.

2. The two regions of the kidney are an outer ___________ and an inner ______________

3. ________% of the plasma that enters the glomerulus is filtered.

D. Matching

____1. glomerular filtration

____2. tubular reabsorption

____3. tubular secretion

____4. urine excretion

a. movement of substances from the peritubular capillary blood into the tubular lumen

b. movement of substances from the glomerular capillary blood into the tubular lumen

c. everything filtered or secreted that is not subsequently reabsorbed

d. movement of substances from the tubular lumen into the peritubular capillary blood

E. Indicate the proper sequence through which fluid flows as it traverses the structures in question by writing the identifying letters in the proper order in the blanks.

1. a. ureter ___ ___ ___ ___ ___
   b. kidney
   c. urethra
   d. bladder
   e. renal pelvis

2. a. efferent arteriole ___ ___ ___ ___ ___ ___
   b. peritubular capillaries
   c. renal artery
   d. glomerulus
   e. afferent arteriole
   f. renal vein

3. a. loop of Henle ___ ___ ___ ___ ___ ___ ___
   b. collecting duct
   c. Bowman's capsule
   d. proximal tubule
   e. renal pelvis
   f. distal tubule
   g. glomerulus

Glomerular Filtration (text page 470)

Contents

The glomerular membrane is more than one-hundred times more permeable than capillaries elsewhere. p. 470

The major force responsible for inducing glomerular filtration is the glomerular-capillary blood pressure. p. 470

The most common factor resulting in a change in GFR is an alteration in the glomerular-capillary blood pressure. p. 472

Glomerular filtration rate can be influenced by changes in the filtration coefficient. p. 477

The kidneys normally receive 20% to 25% of the cardiac output. p. 478

Section Synopsis

Glomerular filtrate is produced as a portion of the plasma flowing through each glomerulus is passively forced under pressure through the glomerular membrane into the lumen of the underlying Bowman's capsule. The net filtration pressure that induces filtration is caused by an imbalance in the physical forces acting across the glomerular membrane. A high glomerular-capillary blood pressure favoring filtration outweighs the combined opposing forces of blood-colloid osmotic pressure and Bowman's-capsule hydrostatic pressure. The glomerular filtration rate (GFR) depends on the net filtration pressure and on the filtration coefficient (K_f). The latter, in turn, depends on the available filtering surface area and on the permeability of the glomerular membrane. The glomerular membrane serves as a sieve to hold back blood cells and plasma proteins while permitting H_2O and other solutes to be pushed through under pressure.

Typically, 20 to 25% of the cardiac output is delivered to the kidneys for the purpose of being acted on by renal regulatory and excretory processes. Of the plasma flowing through the kidneys, normally 20% is filtered through the glomeruli, producing an average GFR of 125 ml/min. This filtrate is identical in composition to plasma except for the plasma proteins that are held back by the glomerular membrane.

The GFR can be altered primarily by changes in the net filtration pressure and to a lesser extent by changes in K_f. The net filtration pressure is altered whenever one of the physical forces acting across the glomerular membrane is changed, of which the glomerular capillary blood pressure is the only factor subject to control. The glomerular capillary pressure depends on the driving force of the mean arterial blood pressure and on the resistance to flow offered by the afferent arterioles. Without any compensatory action, the higher the arterial pressure, the greater the flow of blood into the glomeruli and, accordingly, the higher the glomerular blood pressure and the higher the GFR. At a constant driving pressure, afferent arteriolar vasoconstriction decreases the flow of blood into the glomerulus, resulting in a reduction in glomerular blood pressure and a fall in the GFR. Conversely, afferent arteriolar vasodilation leads to increased glomerular blood flow and a rise in the GFR.

Two mechanisms influence the GFR by adjusting afferent arteriolar caliber: (1) intrinsic renal autoregulation, which keeps the GFR constant despite a change in arterial blood pressure over a range of 80 to 180 mm Hg to prevent inappropriate changes in H_2O, salt, and waste excretion in response to transient changes in blood pressure accompanying daily activities; and (2) extrinsic sympathetic control, which can override autoregulation to deliberately alter the GFR as part of the baroreceptor reflex response to compensate for a change in arterial blood pressure. As the GFR is altered, the amount of fluid lost in the urine is changed correspondingly, providing a mechanism to adjust plasma volume as needed to help restore blood pressure to normal on a long-term basis.

In addition to autoregulatory and sympathetic mechanisms that control the GFR by adjusting the net filtration pressure by means of regulating flow of blood through the afferent arteriole, the GFR can also be regulated by adjustments in the filtration coefficient. The filtration coefficient can be changed by controlling contractile activity within: (1) the mesangial cells within the glomerular tuft, which varies the filtering surface area by opening or closing glomerular capillaries; and (2) Bowman's capsule podocytes, which changes permeability by altering the numer of filtration slits.

Learning Check (Answers on p. A-39)

A. True/False

T/F 1. Part of the kidneys' energy supply is used to accomplish glomerular filtration.

T/F 2. The glomerular capillary wall contains filtration slits formed by the clefts between the foot processes of adjacent podocytes.

T/F 3. The pores in the glomerular membrane are too small for albumin to pass through.

T/F 4. The kidneys receive a disproportionately large share of the cardiac output for the purpose of adjusting and purifying the plasma.

T/F 5. The GFR cannot be altered when the arterial blood pressure is in the range of 80 to 180 mm Hg.

T/F 6. Autoregulation is important to prevent unintentional shifts in the GFR that could lead to dangerous imbalances of fluid, electrolytes, and wastes.

T/F 7. Sympathetic vasoconstriction of the afferent arterioles and a resultant fall in the GFR occur as part of the baroreceptor reflex response when the blood pressure is too low.

B. Fill-in-the-blank

1. GFR = ______________ X ________________.

2. The average GFR is ____________ ml/min.

3. The specialized cells of the ___________________ within the ____________________ detect changes in the rate at which fluid is flowing past them through the tubule. In response they secrete vasoactive chemicals that influence the GFR by making adjustments in the caliber of the __________________ arterioles. This is known as the _________________________ feedback mechanism.

C. Indicate whether the following factor would (a) increase or (b) decrease the GFR, if everything else remained constant, by circling the appropriate letter preceding the question.

a b 1. a rise in Bowman's capsule pressure resulting from ureteral obstruction by a kidney stone

a b 2. a fall in plasma protein concentration resulting from loss of these proteins from a large burned surface

a b 3. a dramatic fall in arterial blood pressure following severe hemorrhage (<80mm Hg)

a b 4. afferent arteriolar vasoconstriction

a b 5. tubulo-glomerular feedback response to a reduction in tubular flow rate

a b 6. myogenic response of an afferent arteriole stretched as a result of an increased driving blood pressure

a b 7. sympathetic activity to the afferent arterioles

a b 8. contraction of mesangial cells

a b 9. contraction of podocytes

Tubular Reabsorption and Tubular Secretion (text page 478)

Contents

Section Synopsis

After a protein-free plasma is filtered through the glomerulus, a variety of tubular forces discriminantly act on the filtered fluid to salvage materials of value to the body while eliminating the useless or harmful materials in the urine. Each substance is handled discretely by the tubules, so even though the concentration of all constituents in the initial glomerular filtrate is identical to their concentration in the plasma, (with the exception of plasma proteins), the concentrations of different constituents are variously altered as the filtered fluid flows through the tubular system. The reabsorptive capacity of the tubular system is tremendous. Over 99% of the filtered plasma is returned to the blood through reabsorption. To be reabsorbed, a substance must pass through both membranes and the entire thickness of a tubular cell, through the interstitial fluid, and across the peritubular capillary wall. If any of the steps in this transepithelial transport process is active, the substance is considered to be actively reabsorbed. All five steps must be passive for passive reabsorption to occur. The major substances actively reabsorbed are Na^+, the principal ECF cation; most other electrolytes; and organic nutrients such as glucose and amino acids. The most important passively reabsorbed substances are Cl^-, H_2O, and urea.

The pivotal event upon which most reabsorptive processes is linked in some way is the active reabsorption of Na^+. An energy-dependent Na^+-K^+ ATPase carrier located in the basolateral membrane of each proximal tubular cell transports Na^+ out of the cells into the lateral spaces between adjacent cells. This induces the net reabsorption of Na^+ from the tubular lumen to the peritubular capillary plasma, the vast majority of which takes place in the proximal tubules. The energy used to supply the Na^+-K^+ ATPase carrier is ultimately responsible for the reabsorption from the proximal tubule of Na^+, glucose, amino acids, Cl^-, H_2O, and urea. Specific Na^+ co-transport carriers located at the luminal border of the proximal tubular cell are driven by the Na^+ concentration gradient to selectively transport glucose or an amino acid from the luminal fluid into the tubular cell, from which they eventually enter the plasma. All of the filtered glucose and amino acids normally are retrieved in this manner from the proximal tubule. Chloride is passively reabsorbed down the electrical gradient established by active Na^+ reabsorption. Water is passively reabsorbed as a result of the localized osmotic gradient created in the lateral spaces by active Na^+ reabsorption. Sixty-five percent of the filtered H_2O is reabsorbed from the proximal tubule in this unregulated fashion. This extensive reabsorption of H_2O increases the concentration of other substances remaining in the tubular fluid, most of which are filtered waste products. The small urea molecules are the only waste products that can passively permeate the tubular membranes. Accordingly, urea is the only waste product partially reabsorbed as a result of this concentration effect, to the extent of about 50% of the filtered urea. The other waste products, failing to be reabsorbed, remain in the urine in highly concentrated form.

The other electrolytes actively reabsorbed by the tubules, such as $PO_4^{\equiv}$ and Ca^{++}, have their own independently functioning carrier systems. Because these carriers as well as the organic nutrient co-transport carriers can become saturated, they exhibit maximal carrier-limited transport capacities, or T_ms. Once the filtered load of an actively reabsorbed substance exceeds the T_m, reabsorption proceeds at a constant maximal rate, with the additional filtered quantity of the substance being excreted in the urine. For the filtered organic nutrients, the T_m far exceeds the normal filtered load so that the role of the kidneys is to conserve and prevent excessive losses of these substances rather than to regulate them. Regulation of the plasma concentration of these organic nutrients depends on other control mechanisms. In contrast, the kidney can regulate many of the electrolytes by virtue of the fact that the reabsorptive capacity (T_m) of the tubule is set so as to reabsorb only that amount of the substance required to maintain the appropriate plasma concentration.

Furthermore, for some of these substances, the set level of reabsorption can be altered through hormonal control. Thus, the kidneys do actively participate in regulating the plasma concentration of Ca^{++}, $PO_4^{\equiv}$ and other plasma electrolytes through precisely controlled reabsorptive mechanisms.

In the special case of Na^+, reabsorption early in the nephron occurs in constant unregulated fashion, but the reabsorption of a small percentage of the filtered Na^+ in the distal and collecting tubules is variable and subject to control. The extent of this controlled Na^+ reabsorption depends primarily on the complex renin-angiotensin-aldosterone mechanism. Since Na^+ and its attendant anion, Cl^-, are the major osmotically active ions in the ECF, the ECF volume is determined by the Na^+ load in the body. In turn, the plasma volume, which reflects the total ECF volume, is important in the long-term determination of arterial blood pressure. Whenever the Na^+ load / ECF volume / plasma volume / arterial blood pressure are below normal, the kidneys secrete renin, an enzymatic hormone that triggers a series of events ultimately leading to increased secretion of aldosterone from the adrenal cortex. Aldosterone increases Na^+ reabsorption from the distal portions of the tubule, thus ameliorating the original reduction in Na^+ / ECF volume / blood pressure.

Similarly, the vast majority of the filtered H_20 undergoes uncontrolled, obligatory reabsorption following the osmotic gradient established by uncontrolled, active Na^+ reabsorption in the proximal tubule. Only a small percentage of H_20 reabsorption in the distal portions of the tubule varies under hormonal control, depending on the state of hydration of the body. Because of the magnitude of filtration, the ability of the kidneys to adjust the reabsorption of even a small percentage of the filtered Na^+ and H_20 represents significant control over the total amount of these materials conserved or excreted daily.

In addition to controlling the rate of excretion of a variety of substances by reabsorbing them to varying extents, the kidney tubules are able to selectively add some substances to the quantity already filtered by means of the process of tubular secretion. Secretion of substances hastens their excretion in the urine. The most important secretory systems are for: (1) H^+, which is important in the regulation of acid-base balance; (2) K^+, which keeps the plasma K^+ concentration at an appropriate level to maintain normal membrane excitability in muscles and nerves; and (3) organic anions and cations, the most significant benefit of which is more efficient elimination of foreign organic compounds from the body.

Thus, the plasma concentration of some substances can be closely regulated by the kidneys through highly discriminatory reabsorptive or secretory mechanisms (for example, Na^+, $PO_4^{\equiv}$, Ca^{++}, H^+, K^+), whereas for other substances the kidneys preserve whatever plasma concentration is established by other regulatory mechanisms (for example, glucose, amino acids). For still other substances, the kidney serves as the major route of elimination (for example, waste products, foreign compounds).

Learning Check (Answers on p. A-39)

A. List the 5 steps involved in transepithelial transport.

1.

2.

3.

4.

5.

B. Multiple choice

1. Reabsorption of which of the following substances is <u>not</u> linked in some way to active Na^+ reabsorption?

   a. glucose
   b. $PO_4^{\equiv}$
   c. H_2O
   d. urea
   e. Cl^-

2. The plasma concentration of which of the following substances is regulated by the kidneys?

   a. plasma proteins
   b. urea
   c. glucose
   d. $PO_4^{\equiv}$
   e. amino acids

3. Which of the following filtered substances is normally <u>not</u> present in the urine at all?

   a. Na^+
   b. $PO_4^{\equiv}$
   c. urea
   d. H^+
   e. glucose

C. Indicate whether the factor in question would ultimately lead to (a) an increase, or (b) a decrease in Na^+ reabsorption by means of the renin-angiotensin-aldosterone mechanism.

a b 1. a precipitous fall in arterial blood pressure as during hemorrhage

a b 2. a reduction in total Na^+ load in the body

a b 3. a reduction in ECF volume

D. Matching

____ 1. directly stimulates Na^+ reabsorption by the distal and collecting tubules

_____2. acted upon by renin

_____3. secreted by the adrenal cortex

_____4. produced by the liver

_____5. acted upon by converting enzyme

_____6. its secretion is directly stimulated by angiotensin II

_____7. stimulates K^+ secretion by the distal and collecting tubules

_____8. secreted from the granular cells of the renal juxaglomerular apparatus

_____9. inhibits Na^+ reabsorption by the renal tubules.

____10. its secretion is directly stimulated by a low plasma K^+ concentration

____11. a potent constrictor of arterioles

a. renin

b. angiotensinogen

c. angiotensin I

d. angiotensin II

e. aldosterone

f. atrial natriuretic peptide

E. True/False

T/F 1. For a substance to be actively reabsorbed, all of the steps of transepithelial transport require energy expenditure.

T/F 2. Sodium reabsorption is under hormonal control throughout the length of the tubule.

T/F 3. Glucose and amino acids are reabsorbed by secondary active transport.

T/F 4. The tubular cells display a T_m for urea.

T/F 5. The renal threshold for glucose is well above the normal plasma glucose concentration, but the renal threshold for $PO_4^{\equiv}$ is equal to the normal plasma $PO_4^{\equiv}$ concentration.

T/F 6. During acidosis, H^+ secretion increases.

T/F 7. A rise in ECF K^+ concentration leads to increased excitability of heart muscle, possibly producing fatal cardiac arrhythmias.

T/F 8. The Na^+ co-transport system in the proximal tubule facilitates elimination of foreign organic compounds from the body.

F. Fill-in-the-blank

1. The energy-dependent step in Na^+ reabsorption involves the _____________________ located at the ____________________ membrane of the tubular cell.

2. If the plasma concentration of substance X is 200 mg/100 ml and the GFR is 125 ml/min, the filtered load of this substance is ________________________.

3. If the T_m for substance X is 200 mg/min, how much of the substance will be reabsorbed at a plasma concentration of 200 mg/100 ml and a GFR of 125 ml/min? ___________________
How much of substance X will be excreted?_____________

4. The plasma concentration of a particular substance at which its T_m is reached and the substance first starts appearing in the urine is known as the ______________________.

5. __________ is the only ion actively reabsorbed in the proximal tubule and actively secreted in the distal and collecting tubule.

6. The _____________________ transforms many foreign organic compounds into ionic form, which facilitates their elimination from the body because such conversion enables them to be secreted by the organic anion transport system.

Urine Excretion and Plasma Clearance (text page 491)

Contents

Section Synopsis

Of the 125 ml/min filtered, normally only 1 ml/min remains in the tubules to be excreted as urine. Only wastes and excess electrolytes not wanted by the body, dissolved in a given volume of H_2O, are left behind to be eliminated in the urine. The excreted material is removed or "cleared" from the plasma, with plasma clearance referring to the volume of plasma being cleared of a particular substance each minute by means of renal activity. Less than the filtered amount of plasma is cleared of substances that undergo reabsorption, whereas greater than the filtered amount of plasma is cleared of substances that undergo secretion.

The kidneys regulate the ECF osmolarity by excreting urine of varying volumes and concentrations to either eliminate or conserve pure H_2O, depending respectively on whether the body has a H_2O excess or deficit. The kidneys are able to produce urine ranging from 0.3 ml/min at 1200 mosm/l to 25 ml/min at 100 mosm/l by reabsorbing variable amounts of H_2O from the distal portions of the nephron. This is made possible by the establishment of a vertical osmotic gradient ranging from 300 to 1200 mosm/l in the medullary interstitial fluid by means of the loop of Henle countercurrent system. Countercurrent flow within the two limbs of the long Henle's loops of juxtamedullary nephrons, coupled with differences in the transport and permeability properties of the limbs, results in the incremental trapping in the medullary interstitial fluid of salt that is actively pumped out by the ascending limb. This accomplishes two things: the establishment of the medullary osmotic gradient and dilution of the tubular fluid to 100 mosm/l as it enters the distal portions of the tubule. Blood flow through the vasa recta preserves this vertical gradient of hypertonicity while supplying the medullary tissue.

As the diluted tubular fluid flows through the distal and collecting tubules, it is surrounded by interstitial fluid of progressively increasing osmolarity. This vertical osmotic gradient to which the tubular fluid is exposed establishes a passive driving force for progressive reabsorption of H_2O, but the actual extent of H_2O reabsorption depends on the amount of vasopressin (antidiuretic hormone) secreted. Vasopressin increases the permeability of the distal and collecting tubules to H_2O; they are impermeable to H_2O in the absence of vasopressin. The amount of vasopressin secreted, in turn, is controlled by two inputs to the hypothalamic neurons that produce this hormone: hypothalamic osmoreceptors, which are the dominant input, and the left atrial baroreceptors. Input from both of these receptors increases vasopressin secretion in the presence of a H_2O deficit (i.e., hypertonicity or hypotension) and inhibits its secretion in response to a H_2O excess (i.e., hypotonicity or hypertension). The resultant adjustment in vasopressin-controlled H_2O reabsorption helps ameliorate the fluid imbalance. In the presence of maximum vasopressin secretion, maximum H_2O is reabsorbed in the late tubular segments so that a small volume of maximally concentrated urine (1200 mosm/l) is excreted. In the absence of vasopressin, no H_2O is reabsorbed late in the distal and collecting tubules so a large volume of dilute urine (100 mosm/l) is excreted at the same concentration as it entered the distal tubule.

Renal failure occurs when less than 25% of the kidney tissue is functional so that the kidneys are no longer able to excrete sufficient urine to adequately maintain homeostasis. Renal failure may be acute, occurring suddenly and perhaps being reversible, or chronic, occurring slowly but irreversibly.

Once formed, urine is propelled by peristaltic contractions through the ureters from the kidneys to the urinary bladder for temporary storage. The bladder can accommodate up to 250 to 400 ml of urine before stretch receptors within its wall initiate the micturition reflex. This reflex causes involuntary emptying of the bladder by simultaneous bladder contraction and opening of both the internal and external urethral sphincters. Micturition can transiently be voluntarily prevented until a more opportune time for bladder evacuation by deliberate tightening of the external sphincter and surrounding pelvic diaphragm.

Learning Check (Answers on p. A-41)

A. Fill-in-the-blank

1. On the average, of the 125 ml/min of plasma filtered, ________________ ml/min is reabsorbed and _____________ ml/min is excreted as urine.

2. The plasma clearance of the harmless foreign compound ____________ is equal to the GFR.

3. The plasma clearance of the organic anion ________________ is equal to the renal plasma flow.

4. __________% of the filtered H_2O is variably reabsorbed under the control of the hormone _____________ in the distal and collecting tubules.

5. The minimum volume of obligatory H_2O loss that must accompany the excretion of wastes each day is _____ ml.

6. The two major inputs to the hypothalamus that govern vasopressin secretion are _____________________ and ________________.

7. Vasopressin is also known as _______________, indicative of its effect on the kidneys.

8. ________________ is increased urinary output of H_2O with little or no increase in excretion of solutes; ____________ refers to increased excretion of both H_2O and solutes.

9. _________________ renal failure has a rapid onset but may be reversible; ____________ renal failure is slow, progressive, and irreversible.

10. ____________% of the renal tissue can adequately perform all excretory and regulatory functions of the kidney.

11. ________________ is the inability to prevent the discharge of urine.

D. True/False

T/F 1. A plasma clearance of 135 ml/min for a substance when the GFR is 125 ml/min indicates that net secretion of the substance occurs.

T/F 2. The osmolarity of the medullary interstitial fluid always equilibrates with the descending limb of the loop of Henle.

T/F 3. The driving force for H_2O reabsorption across all permeable segments of the kidney tubule is an osmotic gradient.

T/F 4. The receptor sites for vasopressin binding are located on the basolateral border, yet the end result is an increase in H_2O permeability of the luminal border of the tubular cells.

T/F 5. In the tubular segments permeable to H_2O, solute reabsorption is always accompanied by comparable H_2O reabsorption.

T/F 6. Solute excretion is always accompanied by comparable H_2O excretion.

T/F 7. Water excretion can occur without comparable solute excretion.

T/F 8. The epithelial lining of the bladder passively stretches to accommodate a larger volume during bladder filling.

T/F 9. One can deliberately prevent urination in spite of the micturition reflex by voluntarily inhibiting the parasympathetic supply to the bladder to halt bladder contraction.

C. Circle the answer that correctly completes the statement.

If a person ingests excess H_2O, the body fluids become (1.hypertonic, hypotonic). In response, vasopressin secretion is (2.stimulated, inhibited). As a result, a (3.small, large) volume of (4. hypertonic, hypotonic) urine is excreted to help ameliorate the fluid imbalance.

D. Indicate what the osmolarity of the tubular fluid is at each of the designated points in a nephron using the following answer code:

a = isotonic (300 mosm/l)

b = hypotonic (100 mosm/l)

c = hypertonic (1200 mosm/l)

d = ranging from hypotonic to hypertonic (100 mosm/l to 1200 mosm/l)

1. Bowman's capsule a b c d
2. end of proximal tubule a b c d
3. tip of Henle's loop of juxtamedullary nephron a b c d
4. end of Henle's loop of juxtamedullary nephron a b c d
5. end of collecting duct a b c d

E. Multiple Choice

1. Which of the following statements concerning the countercurrent system is <u>incorrect</u>?

   a. The loops of Henle of juxtamedullary nephrons are responsible for establishing a vertical osmotic gradient in the interstitial fluid of the renal medulla by countercurrent multiplication.

   b. The active NaCl pump of the ascending limb of Henle's limb can establish a 1200 mosm/l concentration difference between the ascending and descending limbs at any given horizontal level.

   c. By means of countercurrent exchange, the vasa recta preserve the vertical osmotic gradient while supplying blood to the medullary tissue.

   d. The collecting tubules of all nephrons utilize the driving force of the vertical osmotic gradient to accomplish variable H_2O reabsorption under the control of vasopressin, which governs their permeability.

   e. Sodium chloride trapped in the renal medulla contributes to medullary hypertonicity.

2. Which of the following is a possible consequence of renal failure?

   a. anemia
   b. skeletal disturbances
   c. hypertension
   d. uremia
   e. all of the above

3. Which of the following does <u>not</u> occur as part of the micturition reflex?

   a. stimulation of stretch receptors in the bladder wall

   b. parasympathetic - induced contraction of the bladder

   c. stimulation of the motor neurons supplying the external urethral sphincter and pelvic diaphragm

   d. mechanical opening of the internal urethral sphincter as a result of changes in the shape of the contracting bladder

   e. bladder emptying in infants

F. Indicate whether the portion of the tubule in question is permeable or impermeable to the substance in question using the following answer code:

a = permeable

b = impermeable

1. The ascending limb of Henle's loop is _____ to H_2O.

2. The descending limb of Henle's loop is _____ to H_2O.

3. The vasa recta is _____ to salt and _____ to H_2O.

4. The distal and collecting tubules in the absence of vasopressin are _____ to H_2O.

5. The distal and collecting tubules in the presence of vasopressin are _____ to H_2O.

Chapter in Perspective **(text page 508)**

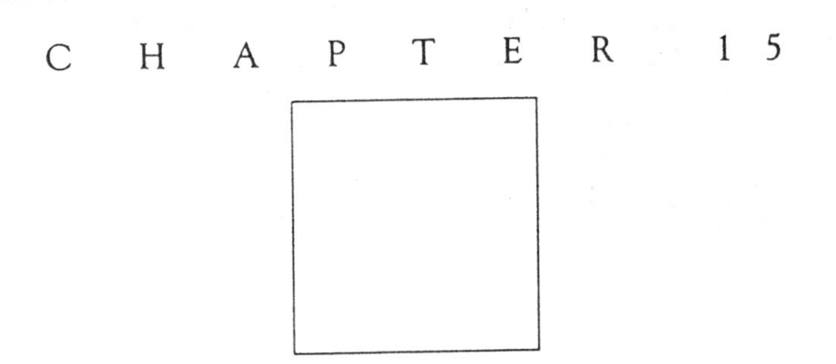

CHAPTER 15

FLUID BALANCE AND ACID-BASE BALANCE

Introduction (text page 510) and Fluid Balance (text page 513)

Contents

Section Synopsis

The body fluids compose 60% of total body weight. This figure varies between individuals, depending on how much fat, a low H_2O content tissue, they possess. Two-thirds of the body H_2O is found in the intracellular fluid (ICF). The remaining one-third present in the extracellular fluid (ECF) is distributed between the plasma (20% of the ECF) and interstitial fluid (80% of the ECF). The small percentage of body H_2O that exists in the lymph and specialized transcellular fluid compartments does not contribute to fluid balance.

Because all plasma constituents are freely exchanged across the capillary walls, the plasma and interstitial fluid are of nearly identical composition, except for the lack of plasma proteins in the interstitial fluid. The interstitial fluid that bathes the body's cells is the true internal environment and, accordingly, its composition must be closely regulated. This regulation is accomplished by mechanisms that act specifically on the plasma but maintain constancy in the entire ECF compartment because the plasma and interstitial fluid are constantly mixing. In contrast, the ECF and ICF have markedly different compositions because the cell membrane barriers are highly selective as to what materials are transported into or out of the cells.

The essential components of fluid balance in the body are maintenance of salt balance and H_2O balance. Input of salt and H_2O into the ECF by way of ingestion must be balanced by output of these materials in order to maintain homeostasis. Control of these factors, in turn, respectively maintains constancy of the ECF volume and ECF osmolarity. Balance of salt and H_2O is maintained primarily by controlling the rate of excretion of these substances in the urine. Control of salt and H_2O intake is of lesser importance and losses of these substances through other avenues is uncontrollable for fluid balance purposes.

Because of the osmotic holding power of Na^+, the major ECF cation, a change in the body's total Na^+ content brings about a corresponding change in ECF volume, including plasma volume, which, in turn, alters the arterial blood pressure in the same direction. Appropriately, changes in ECF volume are primarily monitored by arterial baroreceptors and compensated for in the long run by Na^+-regulating mechanisms. Blood-pressure regulating mechanisms can vary the GFR, and accordingly the amount of Na^+ filtered, by adjusting the caliber of the afferent arterioles supplying the glomeruli. Simultaneously, blood-pressure regulating mechanisms can vary the secretion of aldosterone, the hormone that promotes Na^+ reabsorption by the renal tubules. By varying Na^+ filtration and Na^+ reabsorption, the extent of Na^+ excretion can be adjusted to regulate the plasma volume and subsequently the arterial blood pressure in the long-term.

Changes in ECF osmolarity, in contrast, are primarily detected and corrected by systems responsible for maintaining H_2O balance. The osmolarity of the ECF must be closely regulated to prevent osmotic shifts of H_2O between the ECF and ICF, because cell swelling or shrinking is deleterious, especially to brain neurons. Excess free H_2O in the ECF dilutes the ECF solutes, with the resultant ECF hypotonicity driving H_2O into the cells. An ECF free H_2O deficit, on the other hand, concentrates the ECF solutes. Consequently, H_2O leaves the cells to enter the hypertonic ECF. To prevent these detrimental fluxes, free H_2O balance can be regulated independent of the H_2O that obligatorily follows solutes. Free H_2O balance is accomplished largely by vasopressin and, to a lesser degree, by thirst. Changes in vasopressin secretion and thirst are both governed primarily by hypothalamic osmoreceptors, which monitor ECF osmolarity. The amount of vasopressin secreted determines the extent of free H_2O reabsorption by the distal portions of the nephrons, thereby determining the volume of urinary output. Simultaneously, the intensity of thirst controls the volume of fluid intake. However, because the volume of fluid drunk is often not directly correlated with the intensity of thirst, control of urinary output by means of vasopressin is the most important regulatory mechanism for maintaining H_2O balance.

Learning Check (Answers on p. A-43)

A. True/False

T/F 1. The only avenue by which materials can be exchanged between the cells and the external environment is via the extracellular fluid.

T/F 2. Each intracellular protein exerts more osmotic effect than each intracellular phosphate ion because the proteins are larger.

T/F 3. Diabetes insipidus often leads to hypertonicity of the body fluids.

T/F 4. Water is driven into the cells when the ECF volume is expanded by an isotonic fluid gain.

T/F 5. Salt balance in humans is poorly regulated because of our hedonistic salt appetite.

B. Fill-in-the-blank

1. On the average _______________% of the body weight consists of H_2O.

2. The quantity of any particular substance in the ECF is considered to be a readily available internal____________.

3. When total body input of a particular substance equals its total body output, a _______________ balance exists.

4. The largest body fluid compartment is the ________________.

5. The two components of the ECF compartment are ____________ and _________________.

6. The true internal environment is the _________________.

7. Specialized fluid volumes secreted by specific cells into a particular cavity within the body for a specific purpose are collectively known as ______________________.

8. The barrier between the plasma and interstitial fluid is _____________________.

9. An imbalance in the physical forces, ________________ and ________________, is primarily responsible for producing movement of fluid between the plasma and interstitial fluid.

10. The barrier between the ECF and ICF is the ________________.

11. The force responsible for movement of H_2O between the ECF and ICF is _________________.

12. The principal ECF cation is _________________, which is accompanied primarily by the anion ___________ and to a lesser extent _________________.

13. The major intracellular cation is ________________,

whereas the major intracellular anions are ______________

and ________________.

14. The amount of Na^+ excreted depends on ____________

minus ________________.

15. The amount of Na^+ filtered is controlled by regulating the

________________.

16. The amount of Na^+ reabsorbed is regulated primarily by the

________________ system.

17. The amount of free H_2O reabsorbed is regulated primarily by

the hormone ________________.

C. Multiple Choice

1. Which of the following individuals would have the lowest % of body H_2O?

a. a chubby baby
b. a well-proportioned female college student
c. a well-muscled male college student
d. an obese, elderly woman
e. a lean, elderly man

2. Which of the following factors does <u>not</u> increase vasopressin secretion?

a. ECF hypertonicity
b. alcohol
c. stressful situations
d. an ECF volume deficit
e. angiotensin II

D. 1. List the sources of input and output of H_2O in a daily H_2O balance.

2. Mark an X by the factors that are subject to control to maintain H_2O balance.

3. Mark an * by the most important factor controlling H_2O balance.

Sources of H_2O input	Sources of H_2O output

E. Indicate whether the item in question is referring to
(a) regulation of ECF volume, or
(b) regulation of ECF osmolarity
by circling the appropriate letter.

a b 1. primarily important to prevent fluid shifts between the ECF and ICF

a b 2. primarily important in the long-term regulation of arterial blood pressure

a b 3. depends primarily on Na^+ balance

a b 4. depends primarily on H_2O balance

a b 5. monitored primarily by hypothalamic osmoreceptors

a b 6. monitored primarily by arterial baroreceptors

F. Indicate the changes in ECF and ICF volume and osmolarity that exist after equilibrium has been established during overhydration (e.g., drinking excess H_2O) and indicate the compensations that occur by circling the appropriate item in parentheses.

During overhydration, ECF volume is (normal, increased above normal, decreased below normal) and ECF osmolarity is (isotonic, hypotonic, hypertonic). After the resultant fluid shift has occurred, ICF volume is (normal, increased above normal,decreased below normal) and ICF osmolarity is (isotonic, hypotonic, or hypertonic). As compensatory measures, vasopressin secretion is (increased, decreased, unchanged), resulting in (increased, decreased, no change in) urinary output, and thirst is (increased, decreased, unchanged).

Acid-Base Balance (text page 526)

Contents

Section Synopsis

Acids liberate free hydrogen ions (H^+) into solution; bases bind with free hydrogen ions and remove them from solution. Acid-base balance refers to the regulation of H^+ concentration ($[H^+]$) in the body fluids. To precisely maintain $[H^+]$, input of H^+ must be balanced by equal output. Input includes metabolic production of H^+ within the body and to a small extent dietary acid ingestion. This input must continually be matched with H^+ output by way of urinary excretion of H^+ and by respiratory removal of H^+-generating CO_2. Furthermore, between the time of its generation and elimination, H^+ must be buffered within the body to prevent marked fluctuations in $[H^+]$.

Hydrogen-ion concentration frequently is expressed in terms of pH, which equals the logarithm of $1/[H^+]$. The normal pH of the plasma is 7.4, slightly alkaline compared to neutral H_2O, which has a pH of 7.0. A pH lower than normal (higher $[H^+]$ than normal) is indicative of a state of acidosis. A pH higher than normal (lower $[H^+]$ than normal) characterizes a state of alkalosis. Fluctuations in $[H^+]$ have profound effects on body chemistry, most notably: (1) changes in neuromuscular excitability, with acidosis depressing excitability, especially of the central nervous system, and alkalosis producing overexcitability of both the peripheral and central nervous system; (2) disruption of normal metabolic reactions by altering the structure and function of all enzymes; and (3) alterations in plasma $[K^+]$ brought about by H^+-induced changes in the rate of K^+ elimination by the kidneys.

The primary challenge of controlling acid-base balance is the maintenance of normal plasma alkalinity in the face of continual addition of H^+ to the plasma from ongoing metabolic activity. The three lines of defense to resist changes in $[H^+]$ are: (1) the chemical buffer systems, (2) respiratory control of pH, and (3) renal control of pH.

A chemical buffer system consists of a pair of chemicals involved in a reversible reaction, one that can liberate H^+ and the other that can bind H^+. A buffer pair acts to minimize any changes in pH that occur by acting according to the law of mass action. There are four major buffer pairs in the body fluids. (1) the H_2CO_3:HCO_3^- buffer pair is the major ECF buffer against non-carbonic acids. This buffer system is extremely important because the lungs regulate the H_2CO_3 member of the pair (as reflected by CO_2), and the kidneys regulate HCO_3^-. (2) Protein buffers are the major intracellular buffers and also contribute to ECF buffering. (3) The hemoglobin buffer system buffers H_2CO_3-generated H^+ in transit between the tissues and lungs. (4) The phosphate buffer pair, NaH_2PO_4:Na_2HPO_4, is an important urinary buffer and also contributes to intracellular buffering.

The respiratory system, constituting the second line of defense, normally eliminates the metabolically produced CO_2 so that H_2CO_3 does not accumulate in the body fluids. When the chemical buffers alone have been unable to immediately minimize a pH change, the respiratory system responds within a few minutes by altering its rate of CO_2 removal. An increase in $[H^+]$ arising from non-carbonic acid sources stimulates respiration so that more H_2CO_3-forming CO_2 is blown off, compensating for the acidosis by reducing the generation of H^+ from H_2CO_3. Conversely, a fall in $[H^+]$ depresses respiratory activity so that CO_2 and thus H^+-generating H_2CO_3 can accumulate in the body fluids to compensate for the alkalosis.

The kidneys are the third and most powerful line of defense. They require hours to days to compensate for a deviation in body-fluid pH, but when they respond they do not merely suppress the change in acid-base status as do the buffer systems. The kidneys not only eliminate the normal amount of H^+ produced from non-H_2CO_3 sources, but they can also alter their rate of H^+ removal in response to changes in both non-H_2CO_3 and H_2CO_3 acids. This is unlike the lungs, which can only adjust H^+ generated from H_2CO_3.

Furthermore, inextricably coupled with the renal H^+ secretory mechanism is the kidneys' ability to regulate [HCO_3^-] in the body fluids as well. The kidneys compensate for acidosis by secreting excess H^+ in the urine while conserving and even adding more HCO_3^- to the plasma to expand the HCO_3^- buffer pool. During alkalosis, the kidneys conserve H^+ by not excreting any H^+ in the urine while eliminating the excess HCO_3^-. Secreted H^+ that is to be excreted in the urine must be buffered in the tubular fluid to prevent the H^+ concentration gradient from becoming so great that it prevents further H^+ secretion. Normally H^+ is buffered by the urinary phosphate buffer pair, which is abundant in the tubular fluid because excess dietary phosphate spills into the urine to be excreted from the body. In the face of a pronounced acidosis, when all of the phosphate buffer is already used up in buffering secreted H^+, the kidneys are capable of secreting NH_3 into the tubular fluid to serve as a buffer so that H^+ secretion can continue.

There are four types of acid-base imbalances: respiratory acidosis, respiratory alkalosis, metabolic acidosis, and metabolic alkalosis. Respiratory acid-base disorders originate with deviations from normal in [CO_2], whereas metabolic acid-base imbalances are associated with an unintentional shift in [HCO_3^-]. Through use of the Henderson-Hasselbalch equation, pH can be directly related to the [HCO_3^-]/[CO_2] ratio, which is normally 20/1. Together the kidneys and lungs are important in the long-term maintenance of normal pH because they regulate [HCO_3^-] and [CO_2] respectively. When the ratio falls below 20/1, either because [HCO_3^-] has decreased below normal or [CO_2] has increased above normal, acidosis exists. When the ratio exceeds 20/1, either because [HCO_3^-] is higher than normal or [CO_2] is lower than normal, alkalosis exists.

The three lines of defense participate in correcting the pH imbalance to varying degrees, depending on the original dysfunction responsible for the acid-base problem. To the greatest extent possible, the buffers transiently minimize the shift in pH, whereas in the long run the respiratory system makes compensatory adjustments in [CO_2] and the kidneys make compensatory adjustments in [HCO_3^-]. Compensation involves a deliberate displacement of the normal member of the buffer pair in the same direction as the deviated member so that a normal [HCO_3^-]/[CO_2] ratio can be reestablished and pH can be restored to normal. Respiratory compensation cannot contribute to respiratory acid-base disorders, nor can renal compensation participate in acidosis resulting from kidney failure.

Learning Check (Answers on p. A-44)

A. Fill-in-the-blank

1. ____________ are a special group of hydrogen-containing substances that dissociate in solution to liberate free H^+ and anions.

2. A __________________ is a substance that can combine with a free H^+, thus removing it from solution.

3. A pH of _______ is considered to be chemically neutral. The normal pH of plasma is ______. The pH range compatible with life is ________ to ______.

4. A ________________ is a pair of chemical compounds involved in a reversible reaction, one that can yield free H^+ and one that can bind free H^+, which together minimize changes in pH when either an acid or a base is added to or removed from the solution.

5. Of the two members of the H_2CO_3 : HCO_3^- buffer system, ___[___] is regulated by the lungs whereas ___[___]___ is regulated by the kidneys.

B. True/False

T/F 1. The major source of H^+ generated in the body is via H_2CO_3 formation from metabolically produced CO_2.

T/F 2. Carbon dioxide is unintentionally increased as a cause of respiratory acidosis but is deliberately increased as a compensation for metabolic alkalosis.

T/F 3. Respiratory adjustments can fully compensate for uremic acidosis.

T/F 4. Secreted H^+ that is coupled with HCO_3^- reabsorption is not excreted whereas secreted H^+ that is excreted is linked with the addition of new HCO_3^- to the plasma.

T/F 5. Bicarbonate ions that enter the plasma during HCO_3^- reabsorption are not the same HCO_3^- ions that were filtered.

C. Multiple Choice. Circle all correct answers.

1. pH

   a. equals log $1/[H^+]$

   b. equals pK + log $[CO_2]/[HCO_3^-]$

   c. is high in acidosis

   d. falls lower as $[H^+]$ increases.

   e. is normal when the $[HCO_3^-] / [CO_2]$ ratio is 20/1.

2. Acidosis

   a. causes overexcitability of the nervous system.

   b. exists when the plasma pH falls below 7.35.

   c. occurs when the $[HCO_3^-] / [CO_2]$ ratio exceeds 20/1.

   d. occurs when CO_2 is blown off more rapidly than it is being produced by metabolic activities.

   e. occurs when excessive HCO_3^- is lost from the body such as in diarrhea.

3. In response to the acidosis accompanying diabetes mellitus

a. the $H^+ + HCO_3^- \rightleftarrows H_2CO_3 \rightleftarrows CO_2 + H_2O$ reaction shifts to the right to compensate for the rise in $[H^+]$.

b. ventilation is increased to reduce the concentration of H^+-generating CO_2 in the body fluids.

c. the kidneys secrete more H^+ while conserving and adding new HCO_3^- to the plasma.

d. the kidney tubule cells secrete basic phosphate to buffer secreted H^+ that is to be excreted in the urine.

e. an alkaline urine is produced to compensate for the acidosis.

4. The kidney tubular cells secrete NH_3

a. when the urinary pH becomes too high.

b. when the body is in a state of alkalosis.

c. to enable further renal secretion of H^+ to occur.

d. to buffer excess filtered HCO_3^-

e. when there is excess NH_3 in the body fluids.

D. Matching

____1. buffers H^+ generated from carbonic acid

____2. primary ECF buffer for non-carbonic acids

____3. primary intracellular buffer

____4. important urinary buffer

a. hemoglobin buffer system

b. phosphate buffer system

c. protein buffer system

d. $H_2CO_3 : HCO_3^-$ buffer system

E. Complete the following chart

$\frac{[HCO_3^-]}{[CO_2]}$	What uncompensated abnormality is associated with this ratio?	Which of the following is a possible cause of this abnormality? a = pneumonia b = vomiting c = diabetes mellitus d = anxiety	Given that: log of 10 = 1 log of20 = 1.3 log of40 = 1.6 pK = 6.1, what is the pH with this abnormality?
10/1	1.	2.	3.
20/0.5	4.	5.	6.
20/2	7.	8.	9.
40/1	10.	11.	12.

F. List the three lines of defense against changes in $[H^+]$ and indicate the speed of action of each. Mark an * by the most powerful mechanism.

<u>Lines of Defense</u> <u>Speed of Action</u>

Chapter in Perspective (text page 544)

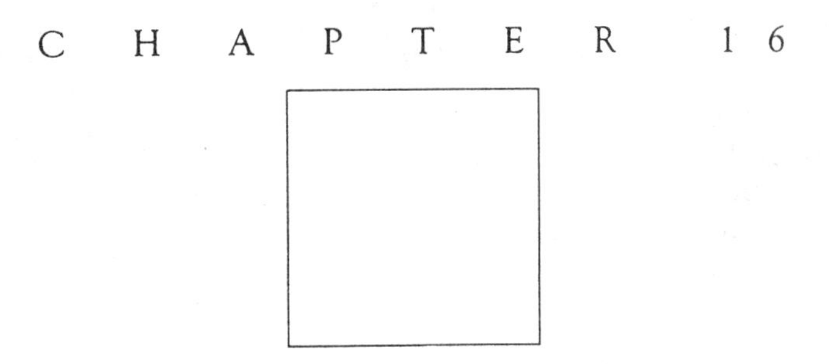

CHAPTER 16

DIGESTIVE SYSTEM

Introduction (text page 546)

Contents

Section Synopsis

The four basic digestive processes are motility, secretion, digestion, and absorption. Digestive activities are carefully regulated by synergistic autonomous, neural (both intrinsic and extrinsic), and hormonal mechanisms to assure that the ingested food is maximally made available to the body for energy production and as synthetic raw materials. The digestive tract consists of a continuous tube that runs from the mouth to the anus, with local modifications that reflect regional specializations for carrying out digestive functions. The lumen of the digestive tract is continuous with the external environment, so its contents are technically outside of the body, a fact that permits digestion of food without committing self-digestion in the process.

Learning Check (Answers on p. A-46)

A. True/False

T/F 1. The extent of nutrient uptake from the digestive tract depends on the body's needs.

T/F 2. The contents of the digestive tract are technically outside of the body.

T/F 3. Slow-wave potentials are always accompanied by contractile activity.

B. Fill-in-the-blank

1. The four basic digestive processes are ________________, ________________, ________________, and ________________.

2. In general, the parasympathetic nervous system is (excitatory or inhibitory) ________________ to the digestive tract, whereas the sympathetic nervous system is ________________.

3. A digestive reflex involving the autonomic nerves is known as a ________________ reflex, whereas a reflex in which all elements of the reflex arc are located within the gut wall is known as a ________________ reflex.

C. List the three categories of foodstuffs and their absorbable units.

category of foodstuff	absorbable unit
1.	
2.	
3.	

D. Matching

_____1. secretes digestive juices

_____2. suspends digestive organs from the inner wall of the abdominal cavity

_____3. secretes a watery lubricating fluid

_____4. absorbs luminal contents

_____5. provides distensibility and elasticity

_____6. constricts the lumen

_____7. shortens the digestive tract

_____8. houses gut-associated lymphoid tissue

_____9. secretes gastrointestinal hormones

_____10. can alter pattern of surface folding

a. serosa

b. inner circular smooth muscle layer

c. mucous membrane

d. lamina propria

e. mesentery

f. muscularis mucosa

g. submucosa

h. outer longitudinal smooth muscle layer

E. Matching

_____1. pathway by which factors outside of the digestive system can influence digestion

_____2. travels via the blood to alter digestive motility and secretion

_____3. responsible for coordinating local activity within the digestive tract

_____4. responsible for establishing the rate of rhythmic contractility

_____5. coordinates activity between different regions of the digestive tract

a. autonomous smooth muscle function

b. extrinsic (autonomic) nerves

c. hormones

d. intrinsic nerve plexuses

Mouth (text page 556); Pharynx and Esophagus (text page 558)

Contents

The oral cavity is the entrance to the digestive tract. p. 556

The teeth accomplish chewing, which breaks up food, mixes it with saliva, and stimulates digestive secretions. p. 556

Saliva begins carbohydrate digestion but plays more important roles in oral hygiene and in facilitating speech. p. 557

The continuous low level of salivary secretion can be increased by simple and conditioned reflexes. p. 557

Digestion in the mouth is minimal, and no absorption of nutrients occurs. p. 558

Swallowing is a sequentially programmed all-or-none reflex. p. 558

During the oropharyngeal stage of swallowing, food is directed into the esophagus and is prevented from entering the wrong passageways. p. 559

The esophagus is guarded by sphincters at both ends. p. 560

Peristaltic waves push the food through the esophagus. p. 560

The gastroesophageal sphincter prevents reflux of gastric contents. p. 561

Esophageal secretion is entirely protective. p. 561

Section Synopsis

Food enters the digestive system through the mouth, where it is chewed and mixed with saliva to facilitate swallowing. The salivary enzyme, amylase, begins the digestion of polysaccharides, a process that continues in the stomach after the food has been swallowed until amylase is eventually inactivated by the acidic gastric juice. More important than its minor digestive function, saliva is essential for articulate speech and plays an important role in dental health. Salivary secretion is controlled by a salivary center in the medulla, mediated by autonomic innervation of the salivary glands.

Following chewing, the tongue propels the bolus of food to the rear of the throat, which initiates the swallowing reflex. The swallowing center in the medulla coordinates a complex group of activities that result in closure of the respiratory passages and propulsion of the food through the pharynx and esophagus into the stomach.

The esophageal secretion, mucus, is protective in nature. No nutrient absorption occurs in the mouth, pharynx, or esophagus.

Learning Check (Answers on p. A-47)

A. Fill-in-the-blank.

1. The _______________ is a common passageway for both the digestive and respiratory systems.
2. The primary wave of peristalsis in the esophagus is initiated by the _______________. If this primary wave fails to push the bolus to the stomach, a secondary peristaltic wave is initiated by the _______________.

B. True/false

T/F 1. Sympathetic stimulation inhibits salivary secretion.

T/F 2. The swallowing center is located in the medulla.

C. Multiple choice

1. Which of the following is (are) accomplished by chewing?

    a. grinding and breaking up food

    b. mixing food with saliva to facilitate swallowing

    c. reflexly increasing salivary, gastric, pancreatic, and bile secretion

    d. Two of the above are correct

    e. All of the above are correct

2. Which of the following is not a function of saliva?

    a. begins digestion of carbohydrate

    b. facilitates absorption of glucose across the oral mucosa

    c. facilitates speech

    d. exerts an antibacterial effect

    e. plays an important role in oral hygiene

D. Matching

_____1. prevents reentry of food into the mouth during swallowing	a. closure of the pharyngoesophageal sphincter
_____2. triggers the swallowing reflex	b. elevation of the uvula
_____3. seals off the nasal passages during swallowing	c. position of the tongue and contraction of muscles in back of the throat
_____4. prevents air from entering the esophagus during breathing	d. closure of the gastroesophageal sphincter
_____5. closes off the respiratory airways during swallowing	e. bolus pushed to the rear of the mouth by the tongue
_____6. prevents gastric contents from backing up into the the esophagus	f. tight apposition of the vocal cords

Stomach (text page 561)

Contents

The body of the stomach does not actively participate in the act of vomiting. p. 565

Gastric pits are the source of gastric digestive secretions. p. 567

Control of gastric secretion involves three phases. p. 569

The stomach lining is protected from gastric secretions by the gastric mucosal barrier. p. 571

Carbohydrate digestion continues in the body of the stomach, whereas protein digestion commences in the antrum. p. 573

The stomach absorbs alcohol and aspirin but no food. p. 573

Section Synopsis

The stomach, a saclike structure located between the esophagus and small intestine, stores ingested food for variable periods of time until the small intestine is ready to further process it for final absorption. The four aspects encompassing gastric motility are gastric filling, storage, mixing, and emptying. Gastric filling is facilitated by plasticity of the gastric smooth muscle and by vagally-mediated receptive relaxation of the stomach musculature. Gastric storage takes place in the body of the stomach, where peristaltic contractions of the thin muscular walls are too weak to mix the contents. Gastric mixing takes place in the thick-muscled antrum as a result of vigorous peristaltic contractions. Gastric emptying is influenced by factors both in the stomach and in the duodenum. The volume and fluidity of chyme in the stomach tend to promote emptying of the stomach contents. The duodenal factors, which are the dominant factors controlling gastric emptying, tend to delay gastric emptying until the duodenum is ready to receive and process more chyme. The specific factors in the duodenum that delay gastric emptying by inhibiting stomach peristaltic activity are the presence of fat, acid, or hypertonic solutions in the duodenal lumen or distention of the duodenal wall.

Gastric motility declines as the meal empties from the stomach and gastric stimulatory factors are withdrawn. After the stomach is finally emptied and the next meal approaches, gastric peristaltic contractions begin once again under vagal influence. Although occurring concurrently with the sensation of hunger, these contractions are not directly responsible for the hunger sensation. Factors unrelated to digestion, such as emotions and pain, can also alter gastric motility through the autonomic nerves. During the act of vomiting, stomach contents are expelled through the mouth by squeezing the relaxed stomach between the diaphragm from above and the abdominal viscera from below as a result of the simultaneous contraction of the diaphragm and abdominal muscles.

Carbohydrate digestion continues in the body of the stomach under the influence of the swallowed salivary amylase; protein digestion is initiated in the antrum of the stomach where vigorous peristaltic contractions mix the food with gastric secretions, converting it to a thick liquid mixture known as chyme. Gastric secretions include: (1) HCl, which activates pepsinogen, denatures protein, and kills bacteria; (2) pepsinogen, which, once activated, initiates protein digestion; (3) mucus, which provides a protective coating to supplement the gastric mucosal barrier, enabling the stomach to contain the harsh luminal contents without self-digestion; (4) intrinsic factor, which plays a vital role in vitamin B_{12} absorption, a constituent essential for normal red blood cell production; and (5) gastrin, a hormone that plays a dominant role in regulating gastric secretion. Histamine, a potent gastric stimulant that is not normally secreted, is released with devastating effects during ulcer formation.

Both gastric motility and gastric secretion are under complex control mechanisms, involving not only gastrin but also vagal and intrinsic nerve responses and enterogastrone hormones secreted from the small intestine. Regulation is aimed at balancing the rate of gastric activity with the ability of the small intestine to handle the arrival of acidic, fat-laden contents from the stomach. No nutrients are absorbed from the stomach.

Learning Check (Answers on p. A-47)

A. True/False

T/F 1. The most important function of the stomach is to begin protein digestion.

T/F 2. Peristaltic waves occur continuously in the stomach at a rate of three contractions/minute, keeping pace with the BER.

T/F 3. The stomach is relaxed during vomiting.

T/F 4. Gastric secretion does not begin until the arrival of food in the stomach.

T/F 5. Acid cannot normally penetrate into or between the cells lining the stomach, which enables the stomach to contain acid without injuring itself.

T/F 6. Individuals with gastric ulcers usually have a high stomach-acid content.

T/F 7. Absorption of foodstuffs commences in the stomach.

B. Fill-in-the-blank

1. When food is broken down and mixed with gastric secretions, the resultant thick liquid mixture is known as ____________.
2. Gastric storage takes place in the ______________ (what part?) of the stomach whereas gastric mixing takes place in the ______________.
3. The most potent stimulus for inhibiting gastric motility and emptying is ____________________________.
4. The three gastrointestinal hormones that function as enterogastrones are ______________, ____________, and ______________.
5. The most potent stimulus for gastrin release is ______________. Gastrin, in turn, is a powerful stimulus for _____________. The PGA is directly inhibited from releasing gastrin by ______________.
6. Carbohydrate digestion takes place in the (what part) ______________ of the stomach under the influence of ______________ (what enzyme), while protein digestion takes place in the ______________ under the influence of ______________.

C. Matching

_____ 1. secrete HCl

_____ 2. secrete gastrin

_____ 3. serve as parent cells of all new cells of the gastric mucosa

_____ 4. secrete pepsinogen

_____ 5. secrete intrinsic factor

a. chief cells

b. parietal cells

c. mucous cells

d. G cells of PGA

D. Matching

_____ 1. activates pepsinogen

_____ 2. inhibits amylase

_____ 3. essential for vitamin B_{12} absorption

_____ 4. can act autocatalytically

_____ 5. not normally secreted but a potent stimulant for acid secretion

_____ 6. breaks down connective tissue and muscle fibers

_____ 7. begins protein digestion

_____ 8. serves as a lubricant

_____ 9. kills ingested bacteria

_____ 10. is alkaline

_____ 11. deficient in pernicious anemia

_____ 12. coats the gastric mucosa

a. pepsin

b. mucus

c. HCl

d. intrinsic factor

e. histamine

Pancreatic and Biliary Secretions (text page 574)

Contents

Section Synopsis

Pancreatic exocrine secretions and bile from the liver both enter the duodenal lumen. Pancreatic secretions include: (1) potent digestive enzymes from the acinar cells, which digest all three categories of foodstuff, and (2) an aqueous Na_2HCO_3 solution from the duct cells, which neutralizes the acidic contents emptied into the duodenum from the stomach. This neutralization is important to protect the duodenum from acid injury and to allow the pancreatic enzymes, which are inactivated by acid, to perform their important digestive functions. Pancreatic secretion is primarily under hormonal control, which matches the composition of the pancreatic juice with the needs in the duodenal lumen.

The liver, the body's largest and most important metabolic organ, performs many varied functions. Its contribution to digestion is the secretion of bile, which contains bile salts. Bile salts aid fat digestion through their detergent action and facilitate fat absorption through formation of water-soluble micelles that can carry the products of fat digestion to their absorptive site. Between meals, bile is stored and concentrated in the gall bladder, which is hormonally stimulated to contract and empty the bile into the duodenum during digestion of a meal. After participating in fat digestion and absorption, bile salts are reabsorbed and returned via the hepatic portal system to the liver, where they not only are resecreted but act as a potent choleretic to stimulate the secretion of even more bile. Bile also contains bilirubin, a derivative of degraded hemoglobin, which is the major excretory product in the feces.

Learning Check (Answers on p. A-48)

A. Matching

_____ 1. activation initiated by enterokinase

_____ 2. only enzyme for fat digestion

_____ 3. similar to digestive enzyme found in saliva

_____ 4. activation initiated by trypsin

a. amylase

b. trypsin

c. chymotrypsin

d. lipase

B. True/false

T/F 1. The pancreas secretes enzymes involved in the digestion of all three categories of foodstuff.

T/F 2. The endocrine pancreas secretes secretin and CCK.

T/F 3. The principal clinical manifestation of pancreatic exocrine insufficiency is incomplete protein digestion resulting from the deficiency of the powerful pancreatic proteolytic enzymes.

T/F 4. The liver receives venous blood coming directly from the digestive tract and arterial blood coming from the aorta.

T/F 5. The gall bladder secretes bile.

T/F 6. In cirrhosis, damaged hepatocytes are permanently replaced by an overgrowth of connective tissue.

C. Fill-in-the-blank

1. The pancreatic _______________ secretion neutralizes acidic gastric contents in the duodenal lumen.

2. Secretin stimulates the pancreatic _______________ cells to secrete _______________, whereas CCK stimulates the pancreatic _______________ cells to secrete _______________.

3. The most potent choleretic is _______________.

4. When worn-out red blood cells are destroyed, the hemoglobin is degraded to a yellow pigment known as _______________, which is excreted into the bile.

5. Excessive accumulation of this pigment in the body produces the condition of _______________.

D. Multiple Choice

1. Bile salts

   a. aid fat digestion through their detergent action

   b. aid fat absorption through micelle formation

   c. are lost in the feces once secreted into the bile

   d. Both a and b above are correct

   e. All of the above are correct

2. Which of the following stimulates gall bladder contraction?

   a. CCK

   b. secretin

   c. sympathetic stimulation

   d. Both a and c above are correct

   e. Both b and c above are correct

3. Which of the following is <u>not</u> a function of the liver?

   a. metabolic processing of carbohydrates, proteins, and fats

   b. secretion of proteolytic digestive enzymes

   c. detoxification and/or degradation of body wastes, hormones, drugs, and foreign compounds

   d. synthesis of plasma proteins essential to the clotting process

   e. storage of glycogen, fats, iron, copper, and many vitamins

Small Intestine (text page 582)

Contents

Section Synopsis

The small intestine is the main site for digestion and absorption. Segmentation, its primary motility, thoroughly mixes the food with pancreatic, biliary, and small intestinal juices to facilitate digestion, plus it exposes the products of digestion to the absorptive surfaces. Between meals, the migrating motility complex sweeps the lumen clean.

The juice secreted by the small intestine does not contain any digestive enzymes. The enzymes synthesized by the small intestine act intracellularly within the brush border membranes of the epithelial cells, completing digestion of carbohydrates and protein before they enter the blood. The energy-dependent process of Na^+ absorption provides the driving force for water, glucose, and amino acid absorption. Fat digestion is accomplished entirely in the lumen of the small intestine by pancreatic lipase. Because of their water insolubility, the products of fat digestion must undergo a series of transformations that enable them to be passively absorbed, eventually entering the lymph. With the exception of iron and calcium, whose absorption is regulated depending on the body's needs, the small intestine absorbs almost everything presented to it, from ingested food to digestive secretions to sloughed epithelial cells. Only a small amount of fluid and nondigestible food residue passes on to the large intestine.

The small intestinal lining is remarkably adapted to its digestive and absorptive function. It is thrown into folds that bear a rich array of fingerlike projections, the villi, which are furnished with a multitude of even smaller hairlike protrusions, the microvilli. Altogether, these surface modifications tremendously increase the area available to house the membrane-bound enzymes and to accomplish both active and passive absorption. This impressive lining is replaced approximately every three days to assure an optimally healthy and functional presence of epithelial cells in spite of harsh luminal conditions.

Learning Check (Answers on p. A-49)

A. True/false

T/F 1. The main function of the ileocecal valve/sphincter is to prevent small intestinal contents from entering the large intestine before digestion of food is completed.

T/F 2. Protein is continually lost from the body through digestive secretions and sloughed epithelial cells, which pass out in the feces.

T/F 3. Most of the fluid presented to the small intestine for absorption has been ingested.

T/F 4. The overall acid-base balance of the body is not normally altered by digestive activities.

B. Fill-in-the-blank

1. The primary mixing and propulsive motility of the small intestine is _______________.

2. The three modifications of the small intestine mucosa that greatly increase the surface area available for absorption are _______________, _______________, and _______________.

3. The entire lining of the small intestine is replaced approximately every _______________ days.

4. The two substances absorbed by specialized transport mechanisms located only in the terminal ileum are _______________ and _______________.

5. The extent of absorption of what two electrolytes is regulated depending on the body's needs? _______________ and _______________.

C. Multiple choice

1. The small intestinal digestive enzymes:

   a. are secreted into the lumen where they perform their function

   b. act intracellularly within the brush borders

   c. complete the digestion of carbohydrates and protein

   d. Both a and c above are correct

   e. Both b and c above are correct

2. Absorption of which of the following substances is linked to active sodium absorption at the basolateral border of the epithelial cell?

   a. water

   b. glucose

   c. galactose

   d. fructose

   e. amino acids

   f. small peptides

   g. monoglycerides and free fatty acids

3. Which of the following sustances will <u>not</u> enter the portal blood upon absorption?

   a. monoglycerides

   b. glucose

   c. sodium

   d. amino acids

   e. bile salts

Large Intestine (text page 596)

Contents

The large intestine is primarily a drying and storage organ. p. 596

The large intestine is thrown into haustral sacs. p. 596

Haustral contractions slowly shuffle the colonic contents back and forth while mass movements propel colonic contents long distances. p. 596

Feces are eliminated by the defecation reflex. p. 598

Constipation occurs when the feces become too dry. p. 598

Large-intestine secretion is protective in nature. p. 598

The large intestine absorbs salt and water, converting the luminal contents into feces. p. 598

Intestinal gases are absorbed or expelled. p. 599

Section Synopsis

The colon serves primarily to concentrate and store undigested food residues and biliary waste products until they can be eliminated from the body as feces. No secretion of digestive enzymes or absorption of nutrients takes place in the colon, all nutrient digestion and absorption having been completed in the small intestine. Haustral contractions slowly shuffle the colonic contents back and forth to accomplish absorption of most of the remaining fluid and electrolytes. Mass movements occur several times a day, usually following meals, propelling the feces long distances. Movement of feces into the rectum triggers the defecation reflex, which can be voluntarily prevented by contraction of the external anal sphincter if the time is inopportune for elimination. If elimination is delayed for too long, constipation may result. The alkaline mucus secretion of the large intestine is primarily protective in nature. Colonic bacteria thrive on undigested food residues, producing some useful products such as vitamins. They also produce the largest source of intestinal gas, most of which is absorbed but some of which is passed to the exterior.

Learning Check (Answers on p. A-50)

A. True/false

T/F 1. Foodstuffs not absorbed by the small intestine are absorbed by the large intestine.

T/F 2. Upon arrival of a new meal in the stomach, the gastrocolic reflex pushes the colonic contents into the rectum, triggering the defecation reflex to eliminate the remains of a preceding meal.

T/F 3. Symptoms associated with constipation are attributable to toxins absorbed from the retained fecal material.

T/F 4. The external anal sphincter is voluntarily relaxed to permit flatus to escape.

B. Fill-in-the-blank

1. _______________ contractions are responsible for mixing the colonic contents, while _______________ periodically propel the contents long distances.

2. Colonic secretion consists of_____________________________.

Overview of the Gastrointestinal Hormones (text page 599)

Chapter in Perspective (text page 600)

C H A P T E R 17

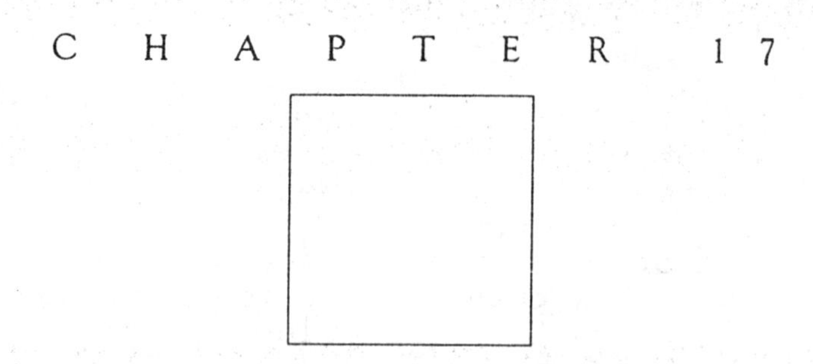

Energy Balance and Temperature Regulation

Introduction (text page 602)

Energy Balance (text page 603)

Contents

Most food energy is ultimately converted into heat in the body. p. 603

The metabolic rate can be measured indirectly by means of the energy equivalent of oxygen. p. 604

Energy input must equal energy output to maintain a neutral energy balance. p. 605

Food intake is controlled primarily by the hypothalamus, but none of the proposals for the mechanism(s) involved are fully satisfactory. p. 606

Obesity occurs when more calories are consumed than are burned up. p. 608

Persons suffering from anorexia nervosa have a pathological fear of gaining weight. p. 609

A Closer Look at Exercise Physiology - All Fat is Not Created Equal p. 610

Section Synopsis

Energy input to the body in the form of food energy must equal energy output, because energy cannot be created or destroyed. Energy expenditure includes: (1) external work performed by skeletal muscles to accomplish movement of an external object or movement of the body through the external environment; and (2) internal work, which consists of all other energy-dependent activities that do not accomplish external work, including active transport, smooth- and cardiac-muscle contraction, glandular secretion, and protein synthesis. Only about 75% of the chemical energy in food is harnessed to do biological work. The rest is immediately converted to heat. Furthermore, all of the energy expended to accomplish internal work is eventually converted into heat. Similarly, 75% of the energy expended by working skeletal muscles is lost as heat. Therefore, most of the energy in food ultimately appears as body heat.

The metabolic rate, which is energy expenditure per unit of time, is measured in kilocalories of heat produced/hour. The basal metabolic rate (BMR) is the rate of energy expenditure determined under prescribed conditions established to identify the basic energy cost of life-sustaining activities, such as pumping blood, breathing, maintaining ionic concentration gradients, and so on. The metabolic rate can be determined directly by a cumbersome method that measures output of heat by the body or, more conveniently, by an indirect method that measures O_2 consumption, then converts that value into rate of heat production using the energy equivalent of O_2, 4.8 kilocalories of heat liberated per one liter of O_2 consumed.

For a neutral energy balance, the energy in ingested food must equal energy expended in performing work. If more food is consumed than energy expended, the extra energy is stored in the body, primarily as adipose tissue, so body weight increases. On the other hand, body energy stores are used to support energy expenditure if more energy is burned than is available in the food, so body weight decreases. Usually body weight remains fairly constant over a prolonged period of time (except during growth) because food intake is adjusted to match energy expenditure on a long-term basis.

Food intake is controlled primarily by the hypothalamus via complex, poorly understood regulatory mechanisms in which hunger and satiety are important components. Among the leading proposals for control of food intake are: (1) the glucostatic theory (increased utilization of glucose signals satiety); (2) the lipostatic theory (increased fat storage signals satiety); (3) the ischymetric theory (increased power production by the cells signals satiety). Surprisingly, the extent of gastrointestinal distention plays little role in determining onset or cessation of eating. Feeding behavior is also influenced by psychosocial factors that can reinforce or override the basic homeostatic regulatory mechanisms controlling food intake.

The development of obesity occurs when food intake exceeds physical activity over a period of time. Once the weight is gained, obese persons tend to maintain their weight constant but at a higher set point. Sufferers of anorexia nervosa, in contrast, have an excessive fear of obesity and a distorted body image that drives them to starve themselves.

Learning Check (Answers on p. A-51)

A. True/False

T/F 1. Energy cannot be created or destroyed, but it can be transformed from one form of energy into another.

T/F 2. If more food energy is consumed than expended, the excess energy is lost as heat.

T/F 3. All of the energy within nutrient molecules can be harnessed to perform biological work.

T/F 4. All energy liberated from ingested food that is not directly used for movement of external objects or stored in the body eventually becomes thermal energy.

T/F 5. Each liter of O_2 contains 4.8 kilocalories of heat energy.

T/F 6. The most dominant factor influencing the BMR is the level of circulating thyroid hormone.

T/F 7. An animal can be driven to overeat either by selective stimulation of the appetite centers or destruction of the satiety centers.

T/F 8. Fat people tend to eat more on a daily basis than thin persons.

B. Fill-in-the-blank.

1. ________________ work refers to the energy expended by contracting skeletal muscles to accomplish movement of objects or the body within the external environment.

2. ________________ work constitutes all forms of biological energy expenditure that do not accomplish mechanical work outside of the body.

3. The ________________ is the energy expenditure (i.e., rate of heat production) per unit of time.

4. The unit of heat energy used in energy balance is the ____________________.

5. The minimal waking rate of energy expenditure determined under standardized conditions is known as the __________.

6. The ____________________ refers to the obligatory increase in metabolic rate incurred during the processing and storage of ingested foods.

7. Regulation of ____________________ is the most important factor in the long-term maintenance of energy balance and body weight.

8. The region of the brain primarily responsible for controlling food intake is the ____________________.

9. The disorder characterized by a pathological fear of gaining weight and a distorted body image is ____________________.

C. Multiple choice.

1. Which of the following is (are) associated with thermoregulators?

    a. Body temperature reflects external environment

    b. Body temperature is regulated to remain constant

    c. Characteristic of humans, mammals and birds

    d. Characteristic of fish, amphibians, reptiles, and invertebrates

2. Which of the following is generally <u>not</u> considered to be a possible satiety signal that leads to the cessation of eating?

    a. Distention of the stomach

    b. Increased glucose utilization

    c. Increased fat storage

    d. Increased power production

Temperature Regulation (text page 609)

Contents

The magnitude of heat loss can be adjusted by varying the flow of blood through the skin. p. 616

The hypothalamus simultaneously coordinates heat-production mechanisms and heat-loss and heat-conservation mechanisms to regulate core temperature homeostatically. p. 616

During a fever, the hypothalamic thermostat is "reset" at an elevated temperature. p. 618

Hyperthermia can occur unrelated to infection. p. 619

The extremes of hyperthermia and hypothermia can be fatal. p. 619

Section Synopsis

The body can be considered as a heat-generating core (internal organs, CNS, and skeletal muscles) surrounded by a shell of variable insulating capacity (the skin). The skin exchanges heat energy with the external environment, with the direction and amount of heat transfer depending on the environmental temperature and the momentary insulating capacity of the shell. The four physical means by which heat is exchanged between the body and external environment are: (1) radiation (net movement of heat energy via electromagnetic waves); (2) conduction (exchange of heat energy by direct contact); (3) convection (transfer of heat energy by means of air currents); and (4) evaporation (extraction of heat energy from the body by the heat-requiring conversion of liquid H_2O into H_2O vapor). Because heat energy moves from warmer to cooler objects, radiation, conduction, and convection can be channels for either heat loss or heat gain, depending respectively upon whether surrounding objects are cooler or warmer than the body surface. Normally they represent avenues for heat loss, along with evaporation due to sweating. The temperature of the skin can be adjusted to vary the rate of heat loss to the external environment as needed to help maintain constancy of the core temperature at about 100°F (equivalent to an oral temperature of 98.6°F).

To prevent serious cellular malfunction, the core temperature must be held constant by continuously balancing heat gain and heat loss despite changes in environmental temperature and variation in internal heat production. This thermoregulatory balance is controlled by the hypothalamus. The hypothalamus is apprised of the temperature status of various regions of the body by warm and cold peripheral thermoreceptors as well as by central thermoreceptors, the most important of which are located in the hypothalamus itself. The primary means of heat gain is heat production by metabolic activity, the biggest contributor being skeletal muscle contraction. Heat loss is adjusted by sweating and by controlling to the greatest extent possible the temperature gradient between the skin and surrounding environment. The latter is accomplished by regulating skin vasomotor activity. Vasoconstriction of the skin vessels reduces the flow of warmed blood through the skin so that skin temperature falls. The layer of cool skin between the core and environment serves to increase the insulative barrier between the warm core and the external air. Conversely, skin vasodilation brings more warmed blood through the skin so that skin temperature approaches the core temperature. This reduces the insulative capacity of the skin.

Upon exposure to cool surroundings, core temperature starts to fall as heat loss is increased as a result of the larger than normal skin-to-air temperature gradient. The hypothalamus responds to reduce the heat loss by inducing skin vasoconstriction while simultaneously increasing heat production through shivering. The latter is characterized by rapid, rhythmic, involuntary heat-generating skeletal muscle contractions. Conversely, in response to a rise in core temperature resulting either from excessive internal heat production accompanying exercise or from excessive heat gain upon exposure to a hot environment, the hypothalamus responds by triggering heat loss mechanisms such as skin vasodilation and sweating. If the environmental temperature is below core temperature, skin vasodilation, by bringing the skin temperature closer to that of the core, increases the skin-to-air temperature gradient so that heat loss can be increased. If this is inadequate, or if the air temperature is warmer than core temperature, sweating is the only involuntary mechanism available to maintain heat balance. In both cold and heat responses, voluntary behavioral contributions also contribute importantly to maintenance of thermal homeostasis.

A fever occurs when endogenous pyrogen released from white blood cells in response to infection raises the hypothalamic set-point. An elevated core temperature develops as the hypothalamus initiates cold-response mechanisms to raise core temperature to the new set-point. Body temperature can also be elevated by other means, in which case the condition is known as hyperthermia. Hyperthermia can range from the normal elevated temperature accompanying sustained exercise to fatal heat stroke that occurs when heat-loss mechanisms are overwhelmed by prolonged exposure to excessive heat. Hypothermia, a reduction in body temperature, may prove fatal as all body systems slow down if prolonged and pronounced enough.

Learning Check (Answers on p. A-52)

A. True/False

T/F 1. Oral temperature is the same as core temperature.

T/F 2. A body temperature greater than 98.6°F is always indicative of a fever.

T/F 3. Core temperature is maintained relatively constant but skin temperature can vary markedly.

T/F 4. The nude body absorbs more radiant heat energy than the body covered in light colored clothing.

T/F 5. Sweat that drips off of the body has no cooling effect.

T/F 6. Production of "goosebumps" in response to cold exposure has no value in regulating body temperature.

T/F 7. The posterior region of the hypothalamus triggers shivering and skin vasoconstriction.

T/F 8. Skin vasoconstriction reduces the temperature of the skin.

T/F 9. The body temperature is normally maintained at an elevated level during sustained exercise.

T/F 10. Profuse sweating occurs during heat stroke as body temperature climbs rapidly upward.

B. Fill-in-the-blank

1. The most important factor determining the extent of evaporation of sweat is the ________________ of the air.

2. The ________________ serves as the body's thermostat.

3. The temperature-sensitive receptors that monitor skin and core temperature are termed ________________________.

4. The primary means of involuntarily increasing heat production is ______________________.

5. Increased heat production independent of muscle contraction is known as ______________________.

6. The amount of heat lost to the environment by radiation and conduction-convection is largely determined by the ______________________ between the skin and the external environment.

7. The only means of heat loss when the environmental temperature exceeds core temperature is ________________.

8. ______________________ refers to a state of collapse resulting from reduced blood pressure brought about as a consequence of overtaxing the heat loss mechanisms.

9. Freezing of exposed tissues produces the condition of ______________________.

10. An elevation in body temperature due to reasons other than an infection is known as ______________________.

C. Multiple choice

1. Which of the following statements concerning heat exchange between the body and external environment is <u>incorrect</u>?

   a. Heat gain is primarily by means of internal heat production.

   b. Radiation serves as a means of heat gain but not of heat loss.

   c. Heat energy always moves down its concentration gradient from warmer to cooler objects.

   d. The temperature gradient between the skin and external air is subject to control.

   e. Very little heat is lost from the body by conduction alone.

2. Which of the following statements concerning fever production is <u>incorrect</u>?

   a. Endogenous pyrogen is released by white blood cells in response to microbial invasion.

   b. The hypothalamic set point is elevated.

   c. The hypothalamus initiates cold-response mechanisms to increase the body temperature.

   d. Prostaglandins appear to mediate the effect.

   e. The hypothalamus is not effective in regulating body temperature during a fever.

D. Indicate by circling the appropriate letter which of the following physiological and behavioral responses occur to restore core temperature to normal when (a) it starts to fall upon cold exposure or (b) it starts to rise upon heat exposure. Furthermore, indicate whether each of these responses represents a means of (c) altering heat gain or (d) altering heat loss by circling the appropriate letter.

1.	shivering	a b	c d
2.	skin vasoconstriction	a b	c d
3.	skin vasodilation	a b	c d
4.	decreased muscle tone	a b	c d
5.	reduced voluntary activity	a b	c d
6.	sweating	a b	c d
7.	increasing the insulative capacity of the skin	a b	c d
8.	hunching over	a b	c d
9.	panting in dogs	a b	c d
10.	putting on light colored clothing	a b	c d
11.	bouncing up and down; hand clapping	a b	c d
12.	erection of hair shafts in furred animals	a b	c d

E. Matching. Indicate the mechanisms of heat transfer being described by filling in the blank with the appropriate item in the right column.

_____ 1. sitting on a cold metal chair

_____ 2. sunbathing on the beach

_____ 3. a gentle breeze

_____ 4. sitting in front of the fireplace

_____ 5. sweating

_____ 6. riding in a car with the windows open

_____ 7. lying on an electric blanket

_____ 8. sitting in a wet bathing suit

_____ 9. fanning yourself

_____10. immersion in cool water

a. radiation

b. conduction

c. convection

d. evaporation

Chapter in Perspective (text page 620)

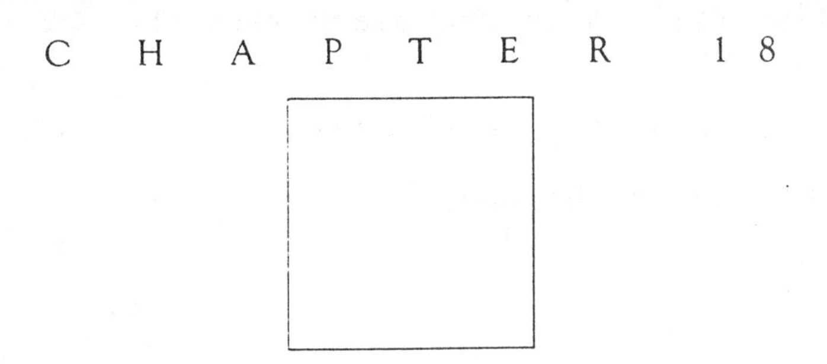

Principles of Endocrinology; Central Endocrine Organs

Introduction (text page 622)

General Principles of Endocrinology (text page 624)

Contents

The effective plasma concentration of a hormone is normally regulated by changes in its rate of secretion. p. 639

The effective plasma concentration of a hormone can be influenced by the hormone's transport, metabolism, and excretion. p. 641

Endocrine disorders are attributable to hormonal excess, hormonal deficiency, or decreased responsiveness of the target tissue. p. 642

The responsiveness of a target tissue to its hormone can be varied by regulating the number of its hormone-specific receptors. p. 643

Section Synopsis

There are four types of intercellular chemical messengers: (1) locally-acting paracrines used by a variety of tissues to influence neighboring cells; (2) locally-acting neurotransmitters used by neurons to rapidly influence the target cells to which they are specifically wired; (3) neurohormones released by neurons into the blood to influence target cells at long distance; (4) and hormones. Hormones are long-distance chemical messengers secreted by the ductless endocrine glands into the blood, which transports them to specific target sites where they regulate or direct a particular function by altering protein activity within the target cells. Hormones do not initiate reactions within their target cells but act by altering the rates at which specific cellular processes proceed. Being diluted in the blood, hormones act at very low plasma concentrations, with their effects being amplified at the target cells.

Even though hormones are able to reach all tissues via the blood, they exert their effects only at their target tissues because of target tissue specificity. Target cells are programmed in two ways to selectively respond to hormones: (1) they are equipped with highly discriminating receptors for binding with only specific hormones, and (2) they are programmed to characteristically respond to a given hormone in a particular way.

In contrast to the other major regulatory system, the rapidly acting nervous system, the endocrine system, controls, by means of its more slowly acting hormonal messengers, activities that require duration rather than speed. The endocrine system is especially important in regulating homeostasis, metabolism, growth, and reproduction. Many complex interactions between the nervous and endocrine systems enable them to work together to coordinate the diverse activities required for maintaining homeostasis.

The means by which hormones are produced by their endocrine glands, are transported to their target site, and interact with their target cells depends on their biochemical structure. The three chemical classes of hormones are peptides, steroids, and amines, the latter including thyroid hormones and adrenomedullary catecholamines. Peptides and catecholamines are hydrophilic; steroid and thyroid hormones are lipophilic. Hydrophilic hormones are synthesized and packaged for export by the endoplasmic reticulum-Golgi apparatus route, stored in secretory vesicles, and released by exocytosis upon appropriate stimulation. They dissolve freely in the blood for transport to their target tissues, where they bind with surface membrane receptors. Upon binding, a hydrophilic hormone triggers a chain of intracellular events by means of a second messenger system that ultimately alters pre-existing cellular proteins, usually enzymes, which exert the effect leading to the target cell's response to the hormone.

Steroids are synthesized by modifications of stored cholesterol by means of enzymes specific for each steroidogenic tissue. Steroids are not stored in the endocrine cell. Being lipophilic, they diffuse out through the lipid membrane barrier as soon as they are synthesized. Control of steroids is directed at their synthesis, not at the release of stored hormone, as is the case with peptides. Thyroid hormone is synthesized and stored in large amounts within extracellular storage pools sequestered "inland" in the thyroid gland. Lipophilic steroids and thyroid hormone are both transported in the blood largely bound to carrier plasma proteins, with only free, unbound hormone being biologically active. The bound hormonal pool serves as a reservoir for the much smaller free effective plasma hormonal pool. Lipophilic hormones readily enter through the lipid membrane barriers of their target cells and bind with intracellular receptors. Steroids bind with cytoplasmic receptors that move, hormone attached, to the nucleus, where they bind with specific sites on the genes. Thyroid hormone binds in the first place with gene receptors in the nucleus. In both instances, hormonal binding activates the synthesis of new intracellular proteins that carry out the hormone's effect on the target cell.

Hormones for the most part are inactivated by the liver and excreted in the urine. The effective plasma concentration of a hormone depends on the balance between its rates of secretion, inactivation, and excretion, and for a lipophilic hormone, its extent of binding to plasma proteins. Generally, the hormonal plasma concentration is controlled by adjustments in the rate of secretion. The primary regulatory inputs influencing the rate of secretion of a particular endocrine cell are: (1) negative-feedback loops; (2) neuroendocrine reflexes; and (3) neurally-controlled endogenous rhythmic oscillations in secretion known as circadian rhythms, which are normally entrained to the light-dark cycle.

Even though target-tissue response is highly dependent on the effective plasma concentration of a hormone, the sensitivity of a target tissue to a given concentration can be modulated by varying the number of receptors available for that particular hormone. The number of receptors can be reduced or "down regulated" by a high concentration of the hormone itself. Furthermore, other hormones can affect a given hormone's receptors, and therefore its effectiveness at its target tissues, through the mechanisms of permissiveness (one hormone increasing the number of receptors for another hormone); synergism (several hormones mutually enhancing each others' receptor activity); and antagonism (one hormone reducing the receptor number for another hormone). In addition, receptors may be lacking or nonresponsive as a result of various disease states.

Endocrine dysfunction arises either from hyposecretion (too little) or hypersecretion (too much) of a hormone or from abnormal tissue responsiveness to a hormone.

Learning Check (Answers on p. A-54)

A. Matching

_____1. hormone — a. local chemical mediator secreted by a variety of cells

_____2. neurohormone — b. local chemical mediator secreted by neurons

_____3. neurotransmitter — c. long distance chemical mediator secreted by endocrine glands

_____4. paracrine — d. long distance chemical mediator secreted by neurons

B. Fill-in-the-blank

1. The specific site upon which a hormone exerts its effect is referred to as a _______________ tissue.
2. A hormone that has as its primary function the regulation of another endocrine gland is classified functionally as a _________________ hormone.
3. The most common site for metabolism of hormones is the ____________________.

4. The primary means of eliminating hormones and their metabolites from the blood is by ________________.

5. Self-induced reduction in the number of receptors for a specific hormone is known as __________________.

6. _________________ refers to enhancement by one hormone of the responsiveness of a target organ to another hormone, for example by means of the first hormone increasing the number of receptors for the second hormone.

C. True/false

T/F 1. One endocrine gland may secrete more than one hormone.

T/F 2. One hormone may influence more than one type of target tissue.

T/F 3. All endocrine glands are exclusively endocrine in function.

T/F 4. A single target tissue may be influenced by more than one hormone.

T/F 5. One target tissue receptor may bind with more than one type of hormone.

T/F 6. Endocrine responses occur more slowly and last longer than neural responses.

T/F 7. Minor differences in structure between hormones within each chemical category often result in profound differences in biological response.

T/F 8. Each steroidogenic organ has all of the enzymes necessary to produce any steroid hormone.

T/F 9. Hormone bound to plasma proteins is not biologically active.

T/F 10. Metabolism of hormones always results in their inactivation.

T/F 11. Hyposecretion or hypersecretion of a specific hormone can occur even though its endocrine gland is perfectly normal.

D. Multiple Choice

1. Which of the following is <u>not</u> controlled at least in part by hormones?

   a. homeostasis

   b. organic metabolism

   c. rapid interactions with the external environment

   d. H_2O and electrolyte balance

   e. adaptation to stress

   f. growth and development

   g. red blood cell production

   h. circulation

   i. digestion and absorption of food

   j. reproduction

2. Which of the following statements concerning control of hormone secretion is <u>not</u> correct?

   a. Normally the effective plasma concentration of a hormone is regulated by appropriate adjustments in the rate of its secretion.

   b. In order to maintain homeostasis, the rate of hormone secretion remains constant.

   c. Negative-feedback control is important in maintaining the plasma concentration of a hormone at a relatively constant set-point.

   d. Neuroendocrine reflexes produce a sudden increase in hormone secretion in response to a specific, usually external, stimulus.

   e. Hormonal secretion fluctuates with time as a result of endogenous oscillators that are entrained to external cues.

E. Use the following answer code to identify the characteristics of various types of hormones.

a = applies only to peptides

b = applies only to steroids

c = applies only to catecholamines

d = applies only to thyroid hormones

e = applies to both peptides and catecholamines

f = applies to both steroids and thyroid hormone

g = applies to both catecholamines and thyroid hormone

h = applies to peptides, catecholamines and thyroid hormone

i = applies to some other combination of hormones

j = applies to all hormones

_____ 1. consist of a chain of specific amino acids of varying length

_____ 2. included in the biochemical category of amines

_____ 3. derived from cholesterol

_____ 4. derived from the amino acid tyrosine

_____ 5. hydrophilic

_____ 6. lipophilic

_____ 7. synthesized and packaged by the endoplasmic reticulum-Golgi apparatus mechanism

_____ 8. formed originally as a large preprohormone molecule

_____ 9. released by exocytosis

_____10. synthesized by cells that internalize low-density lipoproteins (LDL) as raw material

_____11. stored within the endocrine gland following synthesis

_____12. synthesized within an extracellular site in the endocrine gland

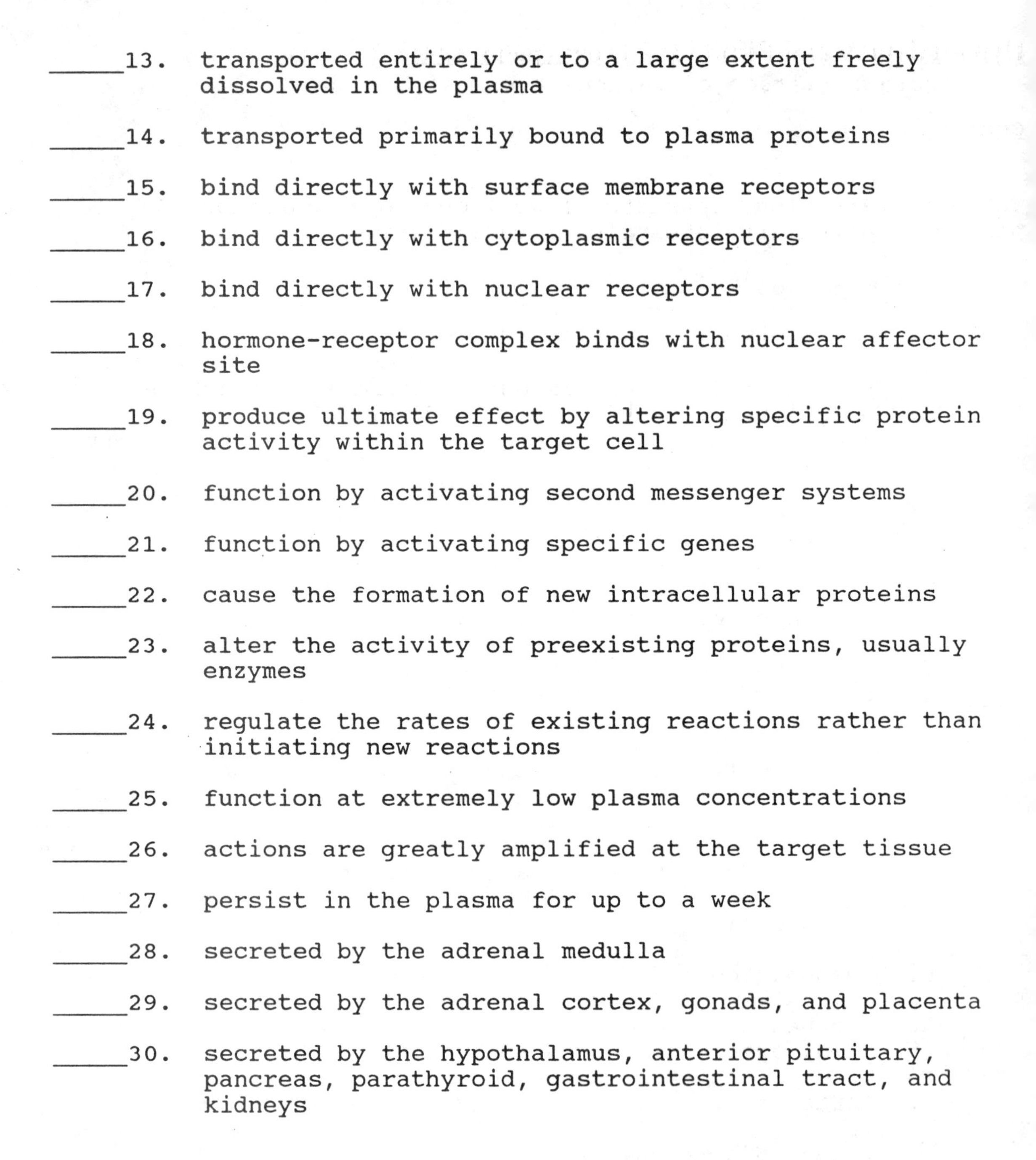

_____13. transported entirely or to a large extent freely dissolved in the plasma

_____14. transported primarily bound to plasma proteins

_____15. bind directly with surface membrane receptors

_____16. bind directly with cytoplasmic receptors

_____17. bind directly with nuclear receptors

_____18. hormone-receptor complex binds with nuclear affector site

_____19. produce ultimate effect by altering specific protein activity within the target cell

_____20. function by activating second messenger systems

_____21. function by activating specific genes

_____22. cause the formation of new intracellular proteins

_____23. alter the activity of preexisting proteins, usually enzymes

_____24. regulate the rates of existing reactions rather than initiating new reactions

_____25. function at extremely low plasma concentrations

_____26. actions are greatly amplified at the target tissue

_____27. persist in the plasma for up to a week

_____28. secreted by the adrenal medulla

_____29. secreted by the adrenal cortex, gonads, and placenta

_____30. secreted by the hypothalamus, anterior pituitary, pancreas, parathyroid, gastrointestinal tract, and kidneys

Hypothalamus and Pituitary (text page 644)

Contents

Section Synopsis

The pituitary gland consists of two distinct lobes, the posterior pituitary and anterior pituitary, which are structurally and functionally connected to the hypothalamus in different ways. The posterior pituitary is essentially a neural extension of the hypothalamus. Collectively the hypothalamus and posterior pituitary form a neurosecretory system for the production, storage, and release of two small peptide hormones, vasopressin and oxytocin. These hormones are synthesized within the cell bodies of neurosecretory neurons located in the hypothalamus, from which they pass down the axon to be stored in nerve terminals within the posterior pituitary. These hormones are independently released from the posterior pituitary into the general circulation in response to action potentials in either vasopressin-secreting or oxytocin-secreting neurosecretory neurons. Generation of action potentials depends on input to the neuronal cell bodies in the hypothalamus, the controlling input being related to the functions performed by the hormones. Vasopressin, whose primary action is related to H_2O retention by the kidneys and consequent expansion cf the plasma volume, is regulated by osmoreceptor and baroreceptor input. Secretion of oxytocin, which exerts important stimulatory effects on the uterus and breasts, appropriately is triggered by stimuli associated with birthing and breastfeeding.

The anterior pituitary, in contrast to the storage function of the posterior pituitary, secretes six different peptide hormones that it produces itself. Five of these hormones are tropic to other endocrine glands (which are designated in parentheses): TSH (thyroid gland); ACTH (adrenal cortex); growth hormone (liver); FSH and LH (gonads). The latter three anterior pituitary hormones exert nontropic effects as well. Growth hormone has important widespread metabolic effects, whereas FSH and LH, known as the gonadotropins, promote gamete (egg and sperm) production. Prolactin, the one anterior pituitary hormone that is completely nontropic, plays a key role in milk production. The anterior pituitary releases its self-synthesized hormones at the bidding of the hypothalamus, which in turn is influenced by a variety of neural and hormonal controlling inputs that it must integrate.

The hypothalamus secretes seven distinct hypophysiotropic hormones into a short vascular portal system, which carries them directly to the anterior pituitary where they either stimulate (hypothalamic releasing-hormone) or inhibit (hypothalamic inhibiting-hormone) the release into the systemic blood of one or more specific anterior pituitary hormones. With each of the hormonal sequences initiated by hypophysiotropic hormones, feedback, usually negative feedback, is exerted on the hypothalamus and/or anterior pituitary by one or more of the hormones in the sequence. The ultimate purpose of a hypothalamic hypophysiotropic hormone - anterior pituitary tropic hormone - target endocrine organ hormone system is served by the final target organ hormone whose plasma concentration must be maintained according to need.

Reviewing the complex patterns of control at each level, negative-feedback mechanisms sustain target-organ hormone concentrations at a prescribed set point, but the set point can be altered by adjustments in the rate of hypophysiotropic hormone secretion according to need as dictated by a variety of controlling inputs to the hypothalamic neurosecretory neurons. The overall regulation of each of the anterior pituitary hormones depends on tropic influences from the hypothalamus ahead of it in the chain of command, and negative-feedback influences from its target organ below it. The final target organ is regulated solely by the appropriate anterior pituitary hormone.

Learning Check (Answers on p. A-56)

A. Indicate which of the following features apply to the posterior and anterior pituitary by circling the appropriate letter, using the answer code below:

a = applies to the posterior pituitary

b = applies to the anterior pituitary

c = applies to both the posterior and anterior pituitary

1. composed of glandular tissue a b c
2. composed of nervous tissue a b c
3. also known as adenohypophysis a b c
4. also known as neurohypophysis a b c
5. secretes MSHs in humans a b c
6. contains pituicytes a b c
7. stores hormones synthesized by the hypothalamus a b c
8. releases hormones into the general circulation a b c
9. its release of hormones is directly controlled by action potentials a b c
10. its release of hormones is directly controlled by hypothalamic hypophysiotropic hormones a b c
11. neurally connected to the hypothalamus a b c
12. connected to the hypothalamus by a vascular link a b c
13. synthesizes the hormones it secretes a b c
14. releases vasopressin and oxytocin into the blood a b c
15. releases primarily tropic hormones into the blood a b c
16. may be directly inhibited by negative feedback from its target organ. a b c

B. Matching

____ 1. stimulates somatomedin secretion by the liver

____ 2. enhances H_2O retention by the kidneys

____ 3. responsible for ovulation

____ 4. stimulates cortisol secretion by the adrenal cortex

____ 5. stimulates testosterone secretion

____ 6. exerts a pressor effect on arterioles

____ 7. stimulate growth of ovarian follicles and development of eggs

____ 8. stimulates uterine contractions

____ 9. regulates overall body growth

____10. stimulates both estrogen and progesterone secretion

____11. stimulates secretion of thyroid hormone

____12. enhances breast development and milk production

____13. promotes milk ejection from the mammary glands

____14. also known as ICSH

____15. important in organic metabolism

____16. required for sperm production

a. vasopressin

b. oxytocin

c. TSH

d. ACTH

e. GH

f. FSH

g. LH

h. prolactin

C. True/False

T/F 1. Melanocyte-stimulating hormones play a role in determining the different amount of melanin in the skin of various human races.

T/F 2. FSH and LH are collectively known as gonadotropins.

T/F 3. Prolactin is the only anterior pituitary hormone that does not exert a tropic action.

T/F 4. Inhibition of the anterior pituitary by a target organ hormone is known as short-loop negative feedback.

D. Multiple choice

1. Which of the following statements concerning hypophysiotropic hormones is (are) correct?

   a. Each hypophysiotropic hormone influences only one anterior pituitary hormone.

   b. All hypophysiotropic hormones stimulate the release of anterior pituitary hormones.

   c. Hypophysiotropic hormones are also produced outside of the hypothalamus, where they serve different functions.

   d. Hypophysiotropic hormones are secreted into the general circulation.

   e. Hypophysiotropic neurons may be inhibited in negative feedback fashion by target organ hormones.

2. Which of the following hormones is <u>not</u> secreted by the hypothalamus?

   a. vasopressin

   b. ACTH

   c. TRH

   d. somatostatin

   e. prolactin-inhibiting hormone

E. Indicate the relationships between the hormones in the hypothalamic/anterior pituitary/adrenal cortex system by using the following answer code to identify which hormone belongs in each blank:

a = cortisol

b = ACTH

c = CRH

(1)_____ from the hypothalamus stimulates the secretion of (2) _______ from the anterior pituitary. (3)____ in turn stimulates the secretion of (4) ____ from the adrenal cortex. In negative feedback fashion, (5) ____ inhibits secretion of (6) ____ and furthermore reduces the sensitivity of the anterior pituitary to (7) ____.

Hormonal Control of Growth **(text page 652)**

Contents

Growth is dependent on growth hormone but is influenced by genetic, environmental, and other hormonal factors as well. p. 653

Growth hormone is essential for growth, but it also exerts metabolic effects not related to growth. p. 654

Growth hormone exerts its growth-promoting effects indirectly by stimulating somatomedins. p. 656

Growth hormone secretion is regulated by two hypophysiotropic hormones. p. 656

Abnormal growth hormone secretion results in aberrant growth patterns. p. 658

Other hormones besides growth hormone are essential for normal growth. p. 659

Section Synopsis

In addition to playing a prominent role in promoting growth, growth hormone exerts metabolic actions unrelated to growth, such as enhancing mobilization of fat and conservation of glucose. Growth is a complex process influenced by genetic, hormonal, and nutritional as well as other environmental factors. Growth is accomplished by lengthening and enlargement of bones, coupled with expansion of soft tissues, as a result of a combination of hyperplasia (increase in the number of cells through cell division) and hypertrophy (increase in the size of cells by means of protein synthesis). Growth hormone promotes growth by stimulating cellular protein synthesis, cell division, and lengthening and widening of bones. In contrast to growth hormone's effects on carbohydrate and fat metabolism, which are exerted directly by this anterior pituitary hormone, growth hormone accomplishes its growth-promoting effects indirectly by stimulating the release of somatomedins. These peptide mediators, which are released into the blood by the liver or act locally at their site of production in the tissues, are responsible for directly exerting the growth-promoting effects dictated by growth hormone.

Growth hormone secretion by the anterior pituitary is regulated in negative-feedback fashion by two hypothalamic hormones, growth hormone releasing-hormone (GHRH) and growth hormone inhibiting-hormone (GHIH or somatostatin). Growth hormone levels are not highly correlated with periods of rapid growth, with the exception of a slight increase during the pubertal growth spurt. Growth hormone levels are no higher during the early childhood growing years than during adult life. The reason for this discrepancy is unclear. The primary signals for increased growth hormone secretion are related to metabolic needs rather than growth; e.g., stress, exercise, low blood glucose levels, high blood amino acid levels, and low fatty acid levels. Growth hormone exhibits a marked diurnal variation in secretion, with levels sharply peaking after the onset of deep sleep.

The most pronounced effects of growth hormone dysfunction are aberrant growth patterns rather than metabolic abnormalities. Other hormones are also essential for normal growth. Thyroid hormone exerts permissive actions for normal growth hormone function. Insulin contributes to growth via its protein-synthesizing effects and its somatomedin-like activity. Androgens play a role in the pubertal growth spurt in both adolescent boys (testicular androgens) and girls (adrenal androgens). Androgens and estrogens both ultimately permanently stop growth by sealing the ends of long bones so that they can no longer respond to growth hormone/somatomedins.

Learning Check (Answers on p. A-57)

A. True/False

T/F 1. Growth hormone exerts metabolic effects unrelated to growth.

T/F 2. Growth hormone levels in the blood are no higher during the early childhood growing years than during adulthood.

T/F 3. Cartilage has no blood supply of its own.

T/F 4. Growth hormone is the most abundant anterior pituitary hormone.

T/F 5. Somatomedins exert insulin-like activity.

B. Fill-in-the-blank

1. ______________________ are the bone cell types that form bone.

2. ______________________ are the bone cell types that dissolve bone.

3. ______________________ are the bone cell types entombed in bone.

4. Activity within the cartilaginous layer of bone known as the ______________________ is responsible for linear growth of the long bones.

5. The growth-promoting actions of growth hormone are directly exerted by peptide mediators known as ______________________.

C. Multiple Choice

Which of the following hormones does <u>not</u> promote growth either directly or indirectly?

1. growth hormone
2. androgens
3. thyroid hormone
4. cortisol
5. insulin

D. Indicate which of the following are associated with growth hormone by circling the appropriate letters.

1. stimulates cell division
2. enhances breakdown of triglyceride fat
3. promotes proliferation of connective tissue
4. promotes glucose uptake by most cells
5. promotes amino acid uptake by most cells
6. stimulates osteoblast activity
7. stimulates protein synthesis and inhibits protein breakdown
8. inhibits chondrocyte activity so that cartilage can be replaced by bone
9. causes the epiphyseal plate to thicken
10. regulated by GHRH and somatostatin
11. secretion markedly increases after onset of deep sleep
12. secretion increases in response to stress and exercise
13. secretion stimulated by a fall in blood glucose
14. secretion stimulated by a fall in blood amino acids
15. secretion stimulated by a fall in blood fatty acids
16. excess secretion commencing in childhood results in acromegaly
17. deficient secretion commencing in childhood results in dwarfism
18. is essential for androgen's growth promoting effects

Chapter in Perspective **(text page 660)**

CHAPTER 19

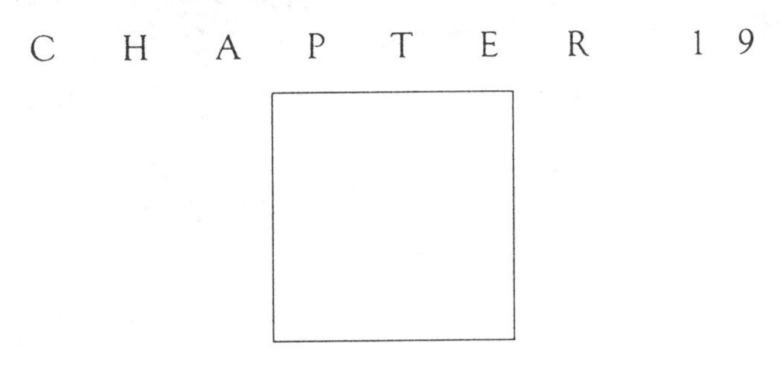

PERIPHERAL ENDOCRINE ORGANS

Introduction (text page 662)

Thyroid Gland (text page 663)

Contents

The major thyroid hormone secretory cells are organized into colloid-filled spheres. p. 663

All of the steps of thyroid hormone synthesis occur on the large thyroglobulin molecule, which subsequently stores the hormones. p. 663

The follicular cells phagocytize thyroglobulin-laden colloid to accomplish thyroid hormone secretion. p. 665

Most of the secreted T_4 is converted into T_3 outside of the thyroid; T_4 and T_3 are both transported largely bound to specific plasma proteins. p. 665

Thyroid hormones are the primary determinants of the body's overall metabolic rate and are also important for bodily growth and for normal development and function of the nervous system. p. 665

Thyroid hormone is regulated by the hypothalamo-pituitary-thyroid axis. p. 666

Abnormalities of thyroid function include both hypothyroidism and hyperthyroidism. p. 667

A goiter may or may not accompany either hypothyroidism or hyperthyroidism. p. 669

Section Synopsis

The bilobed thyroid gland, which lies over the trachea in the neck region, contains two types of endocrine secretory cells: (1) follicular cells, which produce the iodine-containing hormone, T_4 (thyroxine or tetraiodothyronine) and T_3 (triiodothyronine), collectively known as thyroid hormone; and (2) C cells, which synthesize a Ca^{++}-regulating hormone, calcitonin. The control and functions of these thyroid secretory components share little in common other than anatomic proximity.

The follicular cells are arranged into colloid-filled hollow balls known as follicles. Colloid consists of a large glycoprotein, thyroglobulin (TGB), which is produced by the follicular cells and extruded into the lumen of the follicle. All steps of thyroid hormone synthesis take place within the thyroglobulin molecule. Synthesized hormones remain attached to TGB until they are secreted. The following steps are involved in the synthesis, storage, and secretion of thyroid hormone:

(1) The follicular cells actively take up dietary iodine (in its circulating ionized form, iodide) from the blood and oxidize it.

(2) Oxidized iodine binds with the naturally-occurring amino acid tyrosine, which is present on TGB as tyrosyl residues:

1 I + tyrosine → monoiodotyrosine (MIT)
2 I + tyrosine → diiodotyrosine (DIT)

(3) Conjugation of MIT and DIT molecules yield thyroid hormones:

MIT + DIT → triiodothyronine (T_3)
DIT + DIT → tetraiodothyronine (T_4, thyroxine)

(4) Thyroglobulin stores MIT, DIT, T_3 and T_4.

(5) Upon appropriate stimulation, the follicular cells phagocytize a portion of TGB, break it down, release T_3 and T_4 into the blood, deiodinate MIT and DIT, and recycle the freed iodine for further thyroid hormone synthesis.

The most abundant thyroid secretory product is T_4, but T_3 is the most potent. However, most of the secreted T_4 is converted into the more potent T_3 by deiodination outside of the thyroid gland. Both T_3 and T_4 circulate primarily bound to plasma proteins, some of which are specific for thyroid hormone. Only the small unbound portion of the hormone is biologically active.

Thyroid hormone is the primary determinant of overall metabolic rate of the body. By accelerating the metabolic rate of most tissues, it increases O_2 consumption and heat production, the latter being known as its calorigenic effect. It further exerts important influences on many specific reactions involved in intermediary reactions, the net result, especially at high levels, being consumption rather than storage of fuel. This includes degradation of liver glycogen stores, fat stores, and muscle protein. Thyroid hormone is also permissive to the chemical mediators of the sympathetic nervous system. Through this and other means, thyroid hormone indirectly increases cardiac output and increases the work load of the heart. Finally, thyroid hormone is essential for normal growth as well as development and function of the nervous system.

Thyroid hormone secretion is regulated by a long-loop negative feedback system between hypothalamic TRH, anterior pituitary TSH, and thyroid gland T_3 and T_4. The feedback loop maintains thyroid hormone levels relatively constant. Only a few inputs to the hypothalamus are known to be effective in altering TRH and thereby thyroid hormone secretion. Stress inhibits the thyroid axis whereas cold exposure in newborn infants stimulates it.

The thyroid gland can secrete either too little or too much T_3 and T_4, either because of a defect in its stimulatory inputs, a defect in the thyroid gland itself, or a lack of iodine. A goiter (enlarged thyroid) develops when this organ is excessively stimulated. This occurs in some but not all forms of both hypo- and hyperthyroidism. Symptoms of thyroid dysfunction are all directly referable to the consequences of abnormal levels of thyroid hormone on the body.

Learning Check (Answers on p. A-58)

A. True/False

T/F 1. T_3 and T_4 are secreted into the blood by the process of exocytosis.

T/F 2. Thyroid hormone is stored within the colloid attached to thyroglobulin.

T/F 3. MIT and DIT are secreted along with T_3 and T_4.

T/F 4. Most of the secreted T_4 is converted into T_3 outside of the thyroid gland.

T/F 5. Over 99% of the circulating thyroid hormone is bound to plasma proteins.

T/F 6. Thyroid deficient children have stunted growth, which is reversible with thyroid hormone replacement therapy.

T/F 7. Thyroid deficiency from birth results in permanent mental retardation unless replacement therapy is commenced within a few months.

T/F 8. The response to thyroid hormone is detectable within a few minutes after its secretion.

T/F 9. In the absence of TSH, the thyroid gland atrophies.

T/F 10. Hypothyroidism can occur even though the thyroid gland is perfectly normal.

T/F 11. Thyroid-stimulating immunoglobulin stimulates both the secretion and growth of the thyroid similar to TSH.

T/F 12. A goiter is always accompanied by hypersecretion of thyroid hormone.

B. Fill-in-the-Blank

1. The ___________ cells of the thyroid gland secrete the iodine-containing hormones __________ and ____________, whereas the ____________ cells of the thyroid secrete the Ca^{++}-regulating hormone ________________.

2. ____________ and __________ are collectively referred to as thyroid hormone.

3. The lumen of the thyroid follicle is filled with ________, the chief constituent of which is a large, complex glycoprotein known as ____________________.

4. Thyroid hormone is a derivative of the amino acid _____________________.

5. The active transport mechanism for the uptake of iodide from the blood by the thyroid is known as the ____________ or _____________.

6. One I^0 attached to tyrosine yields ____________ whereas the addition of two I^0 atoms to tyrosine forms __________.

7. Coupling of two DITs yields ___________, whereas coupling of one DIT and one MIT results in the formation of __________.

8. The most abundant form of thyroid hormone secreted is ____________, yet ____________ is the most potent thyroid hormone.

9. The three plasma proteins that bind thyroid hormone are ____________, ______________, and ______________.

10. _______________ is the most important physiological regulator of thyroid hormone secretion.

11. An enlarged thyroid is called a _______________.

12. The condition of being hypothyroid from birth is known as __________________.

C. Multiple Choice

1. Which of the following is (are) associated with hyperthyroidism?

   a. poor resistance to cold
   b. myxedema
   c. loss of weight
   d. slow speech, poor memory
   e. exophthalmos

2. Which of the following is <u>not</u> a function of thyroid hormone?

   a. increases the overall metabolic rate

   b. exerts a calorigenic effect

   c. at high levels causes depletion of liver glycogen stores, depletion of fat stores, and muscle wasting as a result of protein degradation.

   d. promotes absorption of dietary iodide by the intestine

   e. causes a proliferation of specific catecholamine target tissue receptors

   f. indirectly leads to an increase in cardiac output

   g. is essential for normal secretion and function of growth hormone

   h. is crucial in the normal development of the nervous system

   i. inhibits the anterior pituitary in negative-feedback fashion

3. Which of the following is <u>not</u> a possible cause of hypothyroidism?

   a. lack of TRH

   b. iodine deficient diet

   c. Grave's disease

   d. lack of an enzyme necessary for one of the steps in the synthesis and release of thyroid hormone

Adrenal Gland and Stress <u>(text page 670)</u>

Contents

The adrenal gland consists of an outer, steroid-secreting adrenal cortex and an inner, catecholamine-secreting adrenal medulla. p. 670

The adrenal cortex secretes mineralocorticoids, glucocorticoids, and sex hormones. p. 670

Mineralocorticoids' major effects are on electrolyte balance and blood pressure-homeostasis. p. 671

Glucocorticoids influence intermediary metabolism and have an important role in adaptation to stress. p. 671

Cortisol secretion is directly regulated by ACTH. p. 672

The adrenal gland may secrete too much or too little of any one of its hormones. p. 674

The catecholamine-secreting adrenal medulla is a modified sympathetic postganglionic neuron. p. 677

Sympathetic stimulation of the adrenal medulla is solely responsible for epinephrine release. p. 680

Adrenomedullary dysfunction is very rare. p. 680

The stress response is a generalized, nonspecific pattern of neural and hormonal reactions to any situation that threatens homeostasis. p. 680

The multifaceted stress response is coordinated within the central nervous system. p. 682

Activation of the stress response by chronic psychosocial stressors may be harmful. p. 683

Section Synopsis

Each of the pair of adrenal glands consists of two separate endocrine organs - an outer steroid-secreting adrenal cortex and an inner catecholamine-secreting adrenal medulla. The adrenal cortex secretes three different categories of steroid hormones. The outer layer, the zona glomerulosa, exclusively secretes mineralocorticoids (primarily aldosterone), whereas the inner two cortical layers, the zona fasiculata and zona reticularis, are the source of glucocorticoids (primarily cortisol) and adrenal sex hormones (primarily the weak androgen, dehydroepiandrosterone). The functional zonation of the adrenal cortex is attributable to regional differences in the concentration of various steroidogenic enzymes responsible for converting the cholesterol precursor molecule into each of the final steroid hormone products.

The mineralocorticoids regulate Na^+ and K^+ balance and are important for blood-pressure homeostasis. The latter is accomplished secondarily as a result of the osmotic effect of Na^+ in maintaining the plasma volume. Because of their ability to conserve Na^+ and thereby maintain an adequate circulating plasma volume, the mineralocorticoids are essential for life. Regulation of mineralocorticoid secretion is largely independent of ACTH. Aldosterone secretion is primarily increased by two mechanisms: (1) activation of the renin-angiotensin-aldosterone mechanism by factors related to a fall in arterial blood pressure; and (2) direct stimulation of the adrenal cortex by a rise in plasma K^+ concentration.

Glucocorticoids help regulate intermediary metabolism and are important in stress adaptation. Specifically, cortisol increases the blood levels of glucose, amino acids, and fatty acids by promoting gluconeogenesis, protein catabolism, and lipolysis. It inhibits glucose uptake by many tissues, thus sparing glucose for use by the glucose-dependent brain. The mobilized amino acids are available for conversion into glucose via hepatic gluconeogenesis or for repair of injured tissues. The glut of fatty acids are used as alternative energy sources for tissues in which glucose uptake is blocked. Cortisol also has extensive permissive actions. At pharmacological doses, it exerts anti-inflammatory and immunosuppressive effects. Cortisol secretion is regulated by a long negative-feedback loop involving hypothalamic CRH and pituitary ACTH. Linked to the sleep-awake cycle, CRH, and accordingly ACTH and cortisol secretion, are subject to a marked diurnal rhythm that peaks in the morning and ebbs at night. The most potent stimulus for increasing activity of the CRH-ACTH-cortisol axis is stress.

The only adrenal sex hormone that has a normal physiological role is dehydroepiandrosterone. It is responsible in females for the sex drive and growth of pubertal hair.

Hypersecretion of any one of the three classes of adrenocortical hormones can occur separately, whereas adrenocortical hyposecretion generally encompasses both mineralocorticoids and glucocorticoids. The symptoms are as would be expected with excess or deficient functions of these hormones.

The adrenal medulla is composed of modified sympathetic postganglionic neurons, which secrete the catecholamine epinephrine into the blood in response to sympathetic stimulation. For the most part, epinephrine reinforces the sympathetic system in its general systemic "fight or flight" responses and in its maintenance of arterial blood pressure. Epinephrine also exerts important metabolic effects, namely increasing blood glucose and blood fatty acids, by promoting glycogenolysis, gluconeogenesis, and lipolysis. The primary stimulus for increased adrenomedullary secretion is activation of the sympathetic system by stress. The only known clinical disease associated with the adrenal medulla is a catecholamine-secreting tumor.

Stress is a generalized, nonspecific response of the body to any factor (a stressor) that threatens maintenance of homeostasis. The stress response involves neurally and hormonally mediated reactions that prepare the body for peak physical responsiveness and protect against anticipated loss of blood. The most prominent components of the stress response are: (1) stimulation of the sympathetic nervous system, which assures adequate delivery of oxygenated blood to tissues most critical for surviving the emergency at the expense of less crucial tissues; (2) enhanced secretion of cortisol, epinephrine, and glucagon, which mobilize the body's energy resources; and (3) activation of the aldosterone pathway and increased vasopressin secretion, which collectively expand the plasma volume and contribute, along with the sympathetic system, to elevation of blood pressure. These multifaceted responses to stress are largely under the direct or indirect control of the hypothalamus. Although these effects are of potential benefit in the face of physical threats, they are largely inappropriate and perhaps detrimental in response to the psychosocial stressors of everyday modern life.

Learning Check (Answers on p. A-59)

A. Fill-in-the-blank

1. The outer portion of the adrenal gland is known as the ____________, which secretes hormones belonging to the chemical class of ____________________.

2. The inner portion of the adrenal gland is known as the ___________________, which secretes hormones belonging to the chemical class of _____________________.

3. The common large precursor molecule that yields ACTH, MSH and ß-endorphin is known as ________________.

4. ___________________ refers to the generalized, nonspecific response of the body to any factor that threatens the maintenance of homeostasis.

B. True/False

T/F 1. "Male" sex hormones are produced in both males and females by the adrenal cortex.

T/F 2. Aldosterone circulates bound primarily to corticosteroid-binding globulin.

T/F 3. Cushing's syndrome is characterized by hyperglycemia, muscle weakness, "moon-face" and "buffalo hump."

T/F 4. Adrenal androgen hypersecretion is usually due to a deficit of an enzyme crucial to cortisol synthesis.

T/F 5. The symptoms of aldosterone hypersecretion include hypernatremia, hyperphosphatemia, and hypertension.

T/F 6. Epinephrine and norepinephrine exert identical effects.

T/F 7. The stress response is believed to be more beneficial for coping with physical threats than psychosocial stressors.

C. Indicate which characteristics apply to the following hormones by circling all appropriate letters using the answer code below. (Note - more than one answer may apply).

a = applies to epinephrine

b = applies to cortisol

c = applies to aldosterone

d = applies to dehydroepiandrosterone

1. produced by the zona glomerulosa a b c d

2. produced by the zona fasiculata and zona reticularis a b c d

3. adrenal androgen a b c d

4. enhances K^+ elimination a b c d

5. stimulates protein degradation a b c d

6. stimulates lipolysis a b c d

7. essential for life a b c d

8. exerts a glucose-sparing effect by inhibiting glucose uptake by many tissues but not the brain a b c d

9. important in the stress response a b c d

10. a catecholamine a b c d

11. a steroid a b c d

12. secretion is stimulated by the sympathetic nervous system a b c d

13. stored in chromaffin granules a b c d

14. promotes female sex drive a b c d

15. secretion is stimulated via the renin-angiotensin system a b c d

16. promotes Na^+ retention a b c d

17. stimulates hepatic gluconeogenesis a b c d

18. stimulates hepatic glycogenolysis a b c d

19. inhibits CRH and ACTH in negative-feedback fashion a b c d

20. secretion is directly stimulated by an increase in plasma K^+ a b c d

21. increases blood glucose levels a b c d

22. increases blood amino acid levels a b c d

23. increases blood fatty acid levels a b c d

24. exerts anti-inflammatory and immunosuppressive effects at pharmacological levels a b c d

25. secretion is largely controlled by ACTH a b c d

26. is secreted in excess in Conn's syndrome a b c d

27. is secreted in excess in Cushing's syndrome a b c d

28. is secreted in excess in adrenogenital syndrome a b c d

29. is secreted in excess by a pheochromocytoma a b c d

30. is deficient in Addison's disease a b c d

31. increases the heart rate a b c d

32. contributes to the "fight or flight" response a b c d

33. displays a marked diurnal rhythm a b c d

34. inhibits secretion of insulin a b c d

D. Listing

1. List the three categories of adrenocortical hormones and name the principal hormone in each category.

Categories of Hormones	Principal Hormone in Category
a.	
b.	
c.	

2. List the hormones important in the stress response.

Endocrine Control of Fuel Metabolism (text page 683)

Contents

Section Synopsis

Intermediary or fuel metabolism refers collectively to the synthesis (anabolism) and breakdown (catabolism) of, as well as interconversions between, the three classes of energy-rich organic nutrients - carbohydrate, fat, and protein - within the body. Glucose and fatty acids derived respectively from carbohydrates and fats are primarily used as metabolic fuels, whereas amino acids derived from proteins are essential for the synthesis of structural and enzymatic proteins. However, amino acids can be converted to glucose and utilized for energy production if necessary, so all three classes of nutrient molecules can be used to provide cellular energy.

There are two metabolic states depending on whether nutrient molecules are being placed in storage or are being mobilized for energy production: the absorptive and postabsorptive state, respectively. During the absorptive state following a meal, the excess absorbed nutrients not immediately needed for energy production or protein synthesis are stored, to a limited extent as glycogen (glycogenesis) in the liver and muscle but mostly as triglycerides (lipogenesis) in adipose tissue. During the postabsorptive state between meals when no new nutrients are entering the blood, the glycogen and triglyceride stores are catabolized (glycogenolysis and lipolysis, respectively) to release nutrient molecules into the blood. If necessary, body proteins are degraded to release amino acids for conversion into glucose (gluconeogenesis). It is essential to maintain the blood glucose concentration above a critical level even during the postabsorptive state, because the brain depends on blood-delivered glucose as its energy source. Accordingly, tissues not dependent on glucose switch to fatty acids as their metabolic fuel, sparing glucose for the brain.

These shifts in metabolic pathways between the absorptive and postabsorptive state are hormonally controlled. The most important hormone in this regard is insulin. Insulin is secreted by the ß cells of the islets of Langerhans, the endocrine portion of the pancreas. The other major pancreatic hormone, glucagon, is secreted by the α cells of the islets. Insulin is an anabolic hormone; it promotes the cellular uptake of glucose, fatty acids, and amino acids and enhances their conversion into glycogen, triglycerides, and proteins, respectively. In doing so, it lowers the blood concentrations of these small organic molecules. Insulin secretion is increased during the absorptive state, primarily by a direct effect of an elevated blood glucose on the ß cells, and is largely responsible for directing the organic traffic into cells during this state. Although insulin promotes energy storage, all of its actions are readily reversible so that a decrease in insulin secretion causes energy mobilization. Insulin secretion is reduced in response to a fall in blood glucose, such as occurs during the postabsorptive state. Thus, changes in insulin secretion are important to fuel metabolism both during a meal (when insulin secretion is increased) and between meals (when insulin levels are low). Glucagon complements the reduced insulin activity during the postabsorptive state to mobilize the energy-rich molecules from their stores. Glucagon, which is secreted in response to a direct effect of a fall in blood glucose on the pancreatic α cells, in general opposes the actions of insulin. By elevating the blood concentrations of glucose and other nutrient molecules (just the opposite of insulin), glucagon is important in orchestrating the metabolic responses that characterize the postabsorptive state. Because insulin and glucagon exert opposite actions and are stimulated by opposite changes in blood glucose, they act in concert as a bihormonal complex for maintaining blood glucose homeostasis.

The primary pancreatic endocrine dysfunction is diabetes mellitus, which encompasses several disorders involving lack of insulin activity. The symptoms are characteristic of an exaggerated postabsorptive state, because insulin is the only hormone that can transfer metabolic fuels into storage and lower their blood concentrations.

Other hormones besides the pancreatic hormones also exert direct metabolic actions, although they do so for reasons other than controlling the feasting-fasting patterns of metabolism. Thyroid hormone accelerates the overall metabolic rate as well as many specific intermediary metabolic reactions. Epinephrine, cortisol, and growth hormone all mobilize energy reserves during stress and exercise. These hormones also play a role in adjustments essential to prolonging survival during long-term starvation.

Learning Check (Answers on p. A-61)

A. True/False

T/F 1. Once a structural protein is synthesized, it remains a part of the cell for the duration of the cell's life.

T/F 2. Excess glucose and amino acids as well as fatty acids can be stored as triglycerides.

T/F 3. Amino acids can be converted to glucose whereas fatty acids cannot.

T/F 4. All forms of diabetes mellitus are characterized by a lack of pancreatic insulin secretion.

T/F 5. Obesity can precipitate overt Type II diabetes mellitus in individuals genetically predisposed.

T/F 6. Both Type I and Type II diabetes must be treated by regular insulin injections.

T/F 7. Reactive hypoglycemia is best treated by a low carbohydrate diet.

T/F 8. Elevated blood amino acid levels stimulate the secretion of both insulin and glucagon even though they exert opposite effects on blood amino acid concentration.

T/F 9. Insulin is the only hormone that can lower blood glucose levels.

B. Fill-in-the-blank

1. The chemical reactions involving the three classes of energy-rich organic molecules are collectively known as ______________________________.

2. ______________________ refers to the synthesis of larger organic molecules from smaller organic molecules.

3. ______________________ refers to the degradation of large energy-rich molecules into smaller organic molecules or into CO_2, H_2O, and energy.

4. The ____________________ is normally dependent on the delivery of adequate blood glucose as its sole source of energy.

5. ______________________ refers to the conversion of glucose into glycogen.

6. ______________________ refers to the conversion of glycogen into glucose.

7. ______________________ refers to the conversion of amino acids into glucose.

8. The ________________ is the principal site for metabolic interconversions of nutrient molecules.

9. Incomplete oxidation of fatty acids yields a group of compounds called ____________________ that can be utilized by the brain for energy during starvation.

10. The three major tissues that are not dependent on insulin for their glucose uptake are ______________, ________________, and ____________________.

C. Multiple Choice

1. Which of the following statements concerning somatostatin is <u>not</u> correct? Somatostatin

   a. is produced by the hypothalamus

   b. is produced by the pancreatic D cells

   c. inhibits growth hormone secretion

   d. inhibits digestion of nutrients and decreased nutrient absorption

   e. is released in response to a fall in blood glucose and blood amino acids

2. Which of the following hormones does <u>not</u> exert a direct metabolic effect?

   a. epinephrine

   b. growth hormone

   c. aldosterone

   d. cortisol

   e. thyroid hormone

D. Indicate the primary circulating form and storage form of each of the three classes of organic nutrients.

	Primary circulating form	Primary storage form
carbohydrate	1.	2.
fat	3.	4.
protein	5.	6.

E. Select from a list of characteristics

1. Indicate which of the following characterize the postabsorptive state.

   a. glycogenolysis
   b. gluconeogenesis
   c. lipolysis
   d. glycogenesis
   e. protein synthesis
   f. triglyceride synthesis
   g. degradation of protein
   h. increased insulin secretion
   i. increased glucagon secretion
   j. glucose-sparing

2. Indicate which of the following are characteristic of insulin.

   a. secreted by pancreatic α cells
   b. secretion stimulated by an increase in blood glucose concentration
   c. lowers blood glucose levels
   d. promotes glycogenesis
   e. promotes gluconeogenesis
   f. facilitates glucose transport into most cells
   g. promotes triglyceride storage
   h. secretion stimulated by an increase in blood amino acid levels
   i. exerts a protein anabolic effect
   j. is present in normal or even elevated levels in Type II diabetes mellitus
   k. is under anterior pituitary control
   l. secretion is stimulated by major gastrointestinal hormones
   m. secretion is stimulated by the sympathetic nervous system
   n. elevates blood amino acid levels

3. Indicate which of the following are associated with diabetes mellitus.

a. hyperglycemia
b. metabolic acidosis
c. glucosuria
d. ketonemia
e. brain starvation
f. polyuria
g. polydipsia
h. polyphagia
i. dehydration
j. coma
k. skeletal muscle weakness
l. blindness
m. gangrenous extremities
n. kidney disease
o. nerve degeneration
p. insulin shock
q. deficient insulin activity
r. elevated glucagon secretion
s. increased incidence of heart disease and atherosclerosis

Endocrine Control of Calcium Metabolism (text page 698)

Contents

Phosphate metabolism is controlled by the same mechanisms that regulate calcium metabolism. p.706

Disorders in calcium metabolism may arise from abnormal levels o parathyroid hormone or vitamin D. p. 707

A Closer Look at Exercise Physiology - Osteoporosis: The Bane of Brittle Bones p. 708

Section Synopsis

Three hormones influence the metabolism of Ca^{++} and $PO_4^{\equiv}$ - parathyroid hormone (PTH), calcitonin, and vitamin D. Plasma Ca^{+} levels are tightly controlled, indicative of the physiologica importance of this ion. Changes in the concentration of free, diffusible plasma Ca^{++}, the biologically active form of this ion, produce profound and life-threatening effects, most notably or neuromuscular excitability. Hypercalcemia reduces excitability whereas hypocalcemia brings about overexcitability of nerves anc muscles, if severe enough to the point of fatal spastic contractions of respiratory muscles.

Parathyroid hormone, whose secretion is directly increased by a fall in plasma Ca^{++} concentration, acts on bone, kidneys, and the intestine to raise the plasma Ca^{++} concentration. In so doing, it is essential for life by preventing the fatal consequences of hypocalcemia. The specific effects of PTH on bone are to promote Ca^{++} efflux from the bone fluid into the plasma across the osteocytic-osteoblastic membrane in the short term and to promote localized dissolution of bone by enhancing activity of the osteoclasts (bone-dissolving cells) in the long term. Dissolution of the $Ca_3(PO_4)_2$-containing bone crystals releases $PO_4^{\equiv}$ as well as Ca^{++} into the plasma. Parathyroid hormone acts on the kidneys to enhance the reabsorption of filtered Ca^{++}, thereby reducing the urinary excretion of Ca^{++} and increasing its plasma concentration. Simultaneously, PTH reduces renal $PO_4^{\equiv}$ reabsorption, in this way increasing $PO_4^{\equiv}$ excretion and lowering plasma $PO_4^{\equiv}$ levels. Thus, $PO_4^{\equiv}$ mobilized by bone dissolution is eliminated from the body while mobilized Ca^{++} is conserved. This is important because a rise in plasma $PO_4^{\equiv}$ would force the deposition of some of the plasma Ca^{++} back into the bone as a result of the inverse relationship between plasma $PO_4^{\equiv}$ and $Ca^{\equiv}$ levels. Furthermore, PTH facilitates the activation of vitamin D, which in turn stimulates Ca^{++} and $PO_4^{\equiv}$ absorption from the intestine. Vitamin D activation is also stimulated by a fall in plasma $PO_4^{\equiv}$ concentration.

Vitamin D can be synthesized from a cholesterol derivative in the skin when exposed to sunlight, but frequently this endogenous source is inadequate because of clothing and indoor dwelling, so vitamin D must be supplemented by dietary intake. From either source, vitamin D must be activated first by the liver and then by the kidneys before it can exert its effect on the intestine. The site of PTH and $PO_4^{\equiv}$ regulation of vitamin D activation is at the kidneys.

Calcitonin, a hormone produced by the C cells of the thyroid gland, is the third factor that regulates Ca^{++}. In negative-feedback fashion, calcitonin is secreted in response to an increase in plasma Ca^{++} concentration and acts to lower plasma Ca^{++} levels by inhibiting activity of bone osteoclasts.

Parathyroid hormone is the most important regulator of minute-to-minute plasma Ca^{++} homeostasis. Both PTH and vitamin D contribute to maintaining the balance between Ca^{++} input and output. Calcitonin is unimportant except during the rare conditions of hypercalcemia. Disorders in Ca^{++} may arise from secretion of too much or too little PTH or from a deficiency of vitamin D.

Learning Check (Answers on p. A-63)

A. True/False

T/F 1. Plasma Ca^{++} concentration is one of the most tightly controlled variables in the body.

T/F 2. The most life-threatening consequence of hypocalcemia is its impact on reducing blood clotting.

T/F 3. All ingested Ca^{++} is indiscriminately absorbed in the intestine.

T/F 4. The greater the physical stress and compression to which a bone is subjected, the greater the rate of bone deposition.

T/F 5. The $Ca_3(PO_4)_2$ bone crystals form a labile pool from which Ca^{++} can rapidly be extracted under the influence of PTH.

T/F 6. An increase in plasma $PO_4^{\equiv}$ concentration forces a reduction in plasma Ca^{++} as a result of the inverse relationship between plasma $PO_4^{\equiv}$ and Ca^{++} levels.

T/F 7. Hyperparathyroidism is characterized by hypercalcemia and hyperphosphatemia.

B. Matching. More than one answer may apply.

____ 1. dissolves bone

____ 2. secretes organic matrix of bone

____ 3. forms a membrane between the bone fluid and ECF across which Ca^{++} efflux occurs

____ 4. stimulated by PTH

____ 5. inhibited by PTH

____ 6. inhibited by calcitonin

____ 7. imprisoned bone cell

a. osteoblast

b. osteocyte

c. osteoclast

C. Matching. More than one answer may apply.

____ 1. important for maintaining constant total amount of Ca^{++} in the body

____ 2. important for maintaining a constant free plasma Ca^{++} concentration

____ 3. PTH contributes to its regulation

____ 4. Vitamin D contributes to its regulation

a. Ca^{++} homeostasis

b. Ca^{++} balance

D. Fill-in-the-blank

1. The ____________________ form of Ca^{++} is biologically active and subject to precise regulation.

2. The three hormonal factors that influence Ca^{++} metabolism are ____________________, ________________, and ________________ .

3. The three compartments with which ECF Ca^{++} is exchanged are ____________________, ________________, and ________________.

4. Bone is a living tissue composed of an organic extracellular matrix impregnated with ________________ crystals consisting primarily of ______________________ salts.

5. Calcitonin is produced by the ___________________ of the ___________________.

6. Vitamin D can be produced by the ______________ upon exposure to sunlight.

7. Vitamin D must be activated by the _________________ and ____________________.

8. The step in vitamin D activation accomplished by the ___________________ is subject to regulation by ___________________ and _______________.

E. Multiple choice

1. Indicate which of the following are characteristics of PTH.

   a. increases plasma Ca^{++} concentration

   b. secretion is increased in response to a rise in plasma Ca^{++} concentration

   c. stimulates Ca^{++} efflux across the osteocytic-osteoblastic membrane

   d. promotes localized dissolution of bone

   e. increases renal reabsorption of Ca^{++}

   f. increases renal reabsorption of $PO_4^{\equiv}$

   g. stimulates activation of vitamin D

   h. directly stimulates Ca^{++} absorption in the intestine

2. Indicate which of the following characterize hypoparathyroidism.

a. osteomalacia

b. osteoporosis

c. rickets

d. hypocalcemia

e. hypophosphatemia

f. overexcitability of the neuromuscular system

g. kidney stones

h. bone demineralization

Chapter in Perspective (text page 709)

CHAPTER 20

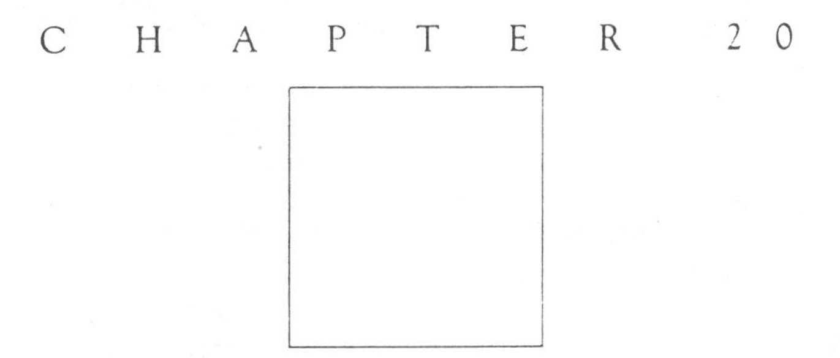

Reproductive Physiology

Introduction (text page 710)

Contents

Section Synopsis

The reproductive system does not contribute to homeostasis and is not essential for survival of an individual, but it is essential for survival of the species. Both sexes produce gametes (reproductive cells), sperm in males and ova (eggs) in females, each of which bears one member of each of the twenty-three pairs of chromosomes present in human cells. Union of a sperm and ovum at fertilization results in the beginning of a new individual with twenty-three complete pairs of chromosomes, half from the father and half from the mother.

The reproductive system is anatomically and functionally distinct in males and females, befitting their different roles in the reproductive process. Males produce sperm and deliver them into the female. Females produce ova, accept sperm delivery, and provide a suitable environment for supporting development of a fertilized ovum until the new individual can survive on its own in the external world. In both sexes, the reproductive system consists of: (1) a pair of gonads, testes in males and ovaries in females, which are the primary reproductive organs that produce the gametes and secrete sex hormones; and (2) a reproductive tract composed of a system of ducts and associated glands that respectively provide a passageway and supportive secretions for the gametes. The externally visible portions of the reproductive system constitute the external genitalia.

Sex determination is a genetic phenomenon dependent on the combination of sex chromosomes at the time of fertilization, an XY combination being a genetic male and an XX combination a genetic female. Sex differentiation refers to the embryonic development of the gonads, reproductive tract, and external genitalia along male or female lines, which gives rise to the apparent anatomic sex of the individual. Early in development, embryoes of both sexes have the same indifferent reproductive tissue capable of differentiating into either a male or female reproductive system, depending on what factors to which it is exposed. In the presence of masculinizing factors, a male reproductive system develops; in their absence, a female system develops. Specifically, an XY embryo bearing testicular determining factor (TDF) within its Y chromosome produces H-Y antigen, which induces the development of testes from the primitive undifferentiated gonadal tissue. The testes, in turn, secrete testosterone and Müllerian-inhibiting factor, which masculinize the entire rest of the reproductive system. On the other hand, an XX embryo, lacking TDF and accordingly H-Y antigen, develops ovaries. Because ovaries do not secrete testosterone or Müllerian-inhibiting factor, female structures develop owing to the lack of male determinants. In most but not all cases, genetic and anatomic sex are compatible.

Learning Check (Answers on p. A-64)

A. Fill-in-the-blank

1. The ______________ are the primary reproductive organs. In males they consist of a pair of ______________ and in females a pair of ___________________.

2. In both sexes the primary reproductive organs produce the reproductive cells or ___________________ which are _________________ in the female and _________________ in the male, as well secrete the sex hormones which are ___________ and ___________________ in the female and __________________ in the male.

3. The reproductive tract includes a system of ducts plus ____________________ that secrete into these passageways.

4. The externally visible portions of the reproductive system are referred to as ______________________.

5. The ________________________ refer to the many external characteristics not directly involved in reproduction that distinguish males and females. The hormone ________________ in males and the hormone _________________ in females are responsible for the development and maintenance of these characteristics.

B. True/False

T/F 1. The reproductive system does not contribute to homeostasis.

T/F 2. Testosterone is directly responsible for differentiating the external genitalia into the penis and scrotum.

T/F 3. It is possible for a genetic male to have the anatomical appearance of a female.

C. Indicate from what primitive reproductive tissue each structure in question differentiates using the answer code below:

a = genital tubercle

b = urethral folds

c = genital swellings

d = Wolffian ducts

e = Müllerian ducts

___ 1. scrotum

___ 2. labia minora

___ 3. glans penis

___ 4. female reproductive tract

___ 5. clitoris

___ 6. penis

___ 7. labia majora

___ 8. male reproductive tract

D. Multiple choice

1. Which of the following statements concerning chromosomal distribution is <u>incorrect</u>?

    a. All human somatic cells contain 23 chromosomal pairs for a total diploid number of 46 chromosomes.

    b. Each gamete contains 23 chromosomes, one member of each chromosomal pair.

    c. During a meiotic reduction division, the members of the chromosome pairs regroup themselves into the original combinations derived from the individual's mother and father for separation into haploid gametes.

    d. Sex determination depends on the combination of sex chromosomes, an XY combination being a genetic male, XX a genetic female.

    e. The sex chromosome content of the fertilizing sperm determines the sex of the offspring.

2. Which of the following statements concerning sex differentiation is <u>incorrect</u>?

    a. A Y chromosome stimulates production of H-Y antigen by primitive gonadal cells.

    b. Ovaries must be present for feminization of the reproductive tract and external genitalia to occur.

    c. H-Y antigen directs differentiation of the primitive gonads into testes.

    d. Secretion of testosterone and Müllerian-inhibiting factor by the fetal testes induces development of the reproductive tract and external genitalia along male lines.

    e. Early in development, embryoes of both sexes have the same indifferent reproductive tissues capable of differentiating into either a male or female reproductive system depending respectively on the presence or absence of masculinizing factors.

Male Reproductive Physiology (text page 718) and

Sexual Intercourse between Males and Females (text page 728)

Contents

Section Synopsis

To be able to reproduce, a male must be capable of producing sperm and depositing them in the female. The male reproductive system consists of the testes, which are the gonads that produce sperm and secrete testosterone, plus a system of ducts and accessory sex glands that respectively deliver sperm to the female and provide supportive secretions for the sperm.

The testes are located outside of the abdominal cavity in the scrotal sac, a site that provides a cooler environment for temperature-dependent spermatogenesis to take place. Spermatogenesis occurs in the highly coiled seminiferous tubules within the testes. Leydig cells located in the interstitial spaces between these tubules secrete the male sex hormone testosterone into the blood. Testosterone is secreted before birth to masculinize the developing reproductive system, then ceases until puberty, at which time it begins once again and continues throughout life. It is responsible for: (1) maturation and maintenance of the entire male reproductive tract; (2) libido; and (3) maturational changes unrelated to reproduction at puberty, including development and maintenance of secondary sexual characteristics.

The testes are regulated by the anterior pituitary hormones, luteinizing hormone (LH) and follicle-stimulating hormone (FSH). These gonadotropic hormones, in turn, are under control of pulsatile secretions of hypothalamic gonadotropin-releasing hormone (GnRH), which begin at puberty for unclear reasons. Testosterone secretion is regulated by LH stimulation of the Leydig cells and, in negative-feedback fashion, testosterone inhibits gonadotropin secretion. Spermatogenesis requires both testosterone and FSH. Testosterone stimulates the mitotic and meiotic divisions required to transform the undifferentiated diploid germ cells, the spermatogonia, into undifferentiated haploid spermatids. Spermatogonia lying at the outer perimeter of the seminiferous tubules divide on a regular basis to provide an ongoing supply of germ cells. Following each mitotic division, one daughter cell remains at the periphery to maintain the germ cell line while the other moves toward the lumen, further dividing and differentiating into spermatozoa. Each developing spermatogonium is capable of yielding sixteen spermatozoa. The remodeling of spermatids into highly specialized motile spermatozoa is stimulated by FSH. A spermatozoon consists only of a DNA-packed head bearing an enzyme-filled acrosome at its tip for penetrating the ovum, a midpiece containing the metabolic machinery for energy production, and a whiplike motile tail. Testosterone and FSH both act via the Sertoli cells, which protect, nurse, and enhance the germ cells throughout their development. Sertoli cells also secrete inhibin, a hormone that inhibits FSH secretion, thus completing the negative-feedback loop.

The still immature sperm are flushed out of the seminiferous tubules into the epididymis by fluid secreted by the Sertoli cells.

The epididymis and ductus deferens store and concentrate the sperm and increase their motility and fertility prior to ejaculation. During ejaculation, the sperm are mixed with secretions released by the accessory glands, which contribute the bulk of the semen. The seminal vesicles supply fructose for energy and prostaglandins, which promote smooth muscle motility in both the male and female reproductive tract to enhance sperm transport. The prostate gland contributes an alkaline fluid for neutralizing the acidic vaginal secretions. The bulbourethral glands release lubricating mucus.

The male sex act consists of erection and ejaculation, which are part of a much broader systemic, emotional reponse that typifies the male sexual response cycle. Erection and ejaculation are both accomplished by spinal reflexes that are triggered by physical or psychic stimuli. Erection is a hardening of the normally flaccid penis, enabling it to penetrate the female vagina. Erection is accomplished by marked vasocongestion of the penis brought about by parasympathetically-induced vasodilation of the arterioles supplying the penile erectile tissue. When sexual excitation reaches a critical peak, ejaculation occurs, which consists of two stages: (1) emission, the emptying of sperm and accessory sex gland secretions into the urethra as a result of sympathetically-induced contraction of the duct walls; and (2) expulsion of semen from the penis as a result of rhythmic contractions of skeletal muscles at the base of the penis. The latter is accompanied by a set of characteristic systemic responses and intense pleasure referred to as orgasm. Following orgasm, the body is gradually returned to its prearousal state.

Females experience a sexual cycle similar to males, with both having excitation, plateau, orgasmic, and resolution phases. The major differences are that women do not ejaculate and they are capable of multiple orgasms. During the female sexual response, the outer third of the vagina constricts to grip the penis while the inner two-thirds expands to create space for sperm deposition.

Learning Check (Answers on p. A-64)

A. True/False

T/F 1. The scrotal location of the testes provides the cooler environment essential for spermatogenesis

T/F 2. Spermatogonia line the lumen of the seminiferous tubule.

T/F 3. Chromosomal replication does not occur during the second meiotic division.

T/F 4. The developing sperm cells arising from a single primary spermatocyte remain joined by cytoplasmic bridges until development is complete so that a Y-bearing sperm can be provided with essential products coded for by the X chromosome.

T/F 5. Testosterone secretion essentially ceases from birth until puberty.

T/F 6. In a sexually mature individual, pulses of GnRH are secreted once a day.

T/F 7. At puberty the hypothalamus becomes more sensitive to feedback inhibition by testosterone.

T/F 8. The pineal gland secretes more melatonin during the light than during the dark.

T/F 9. Although discovered in the semen, prostaglandins are ubiquitous locally-acting chemical messengers that exert widespread effects throughout the body.

T/F 10. Erection is accomplished by contraction of skeletal muscles at the base of the penis.

T/F 11. Females do not experience erection.

T/F 12. Orgasm is very similar in males and females with the exception of no ejaculation in females.

T/F 13. Erection and ejaculation are both spinal reflexes.

T/F 14. During ejaculation the sphincter at the neck of the bladder is opened to permit entry of sperm into the urethra.

T/F 15. Parasympathetic stimulation is required for erection whereas sympathetic stimulation is necessary for ejaculation.

T/F 16. During the female sexual response, the outer third of the vagina expands to accommodate entry of the penis.

T/F 17. Most of the lubrication for sexual intercouse is provided by the female.

B. Fill-in-the-blank.

1. Spermatogenesis takes place within the ____________________ of the testes, stimulated by the hormones ____________________ and ____________________.

2. The ____________________ of the testes secrete testosterone under the hormonal influence of ____________________.

3. An undescended testis is known as ____________________.

4. The ____________________ of a spermatozoon contains the sperm's genetic information; the ____________________ is an enzyme-filled vesicle for penetrating an ovum; power-generating mitochondria are concentrated in the sperm's ____________________; the ____________________ is a whiplike structure comprised of microtubules.

5. The four phases of the sexual cycle in both sexes are ____________________, ____________________, ____________________ and ____________________ phases.

6. The entire process of movement of the sperm and sex gland secretions out of the male reproductive tract is called ____________________; emptying of semen into the urethra is known as ____________________; forceful ejection of sperm from the penis is called ____________________.

C. Matching

____ 1.	secrete(s) prostaglandins		a.	epididymis and ductus deferens
____ 2.	increase(s) motility and fertility of sperm		b.	prostate gland
____ 3.	secrete(s) an alkaline fluid		c.	seminal vesicles
____ 4.	provide(s) fructose		d.	bulbourethral glands
____ 5.	storage site for sperm		e.	penis
____ 6.	concentrate(s) the sperm 100-fold			
____ 7.	secrete(s) fibrinogen			
____ 8.	provide(s) clotting enzymes			
____ 9.	contain(s) erectile tissue			

D. Indicate the chromosomal composition of each of the following stages of developing sperm by circling the appropriate letter using the answer code below:

a = contains a full set of 23 pairs of chromosomes

b = contains a full set of 23 pairs of doubled chromosomes

c = contains a half set of doubled chromosomes

d = contains a set of 23 single unpaired chromosomes

1.	spermatogonium	a	b	c	d
2.	primary spermatocyte	a	b	c	d
3.	secondary spermatocyte	a	b	c	d
4.	spermatid	a	b	c	d
5.	spermatozoan	a	b	c	d

E. Multiple choice

1. Which of the following is (are) <u>not</u> stimulated by testosterone?

   a. masculinization of the developing reproductive tract and external genitalia

   b. descent of the testes into the scrotum

   c. spermatid remodeling

   d. development and maintenance of male secondary sexual characteristics

   e. libido

   f. protein anabolism

   g. bone growth

   h. closure of the epiphyseal plates

   i. mitosis and meiosis of developing sperm cells.

2. Which of the following is (are) <u>not</u> a function of Sertoli cells?

   a. form a blood-testes barrier

   b. phagocytize cytoplasm extruded from sperm during their remodeling

   c. secrete seminiferous tubule fluid

   d. secrete androgen-binding protein

   e. provide nourishment for developing sperm

   f. provide binding sites for LH

   g. secrete inhibin

Female Reproductive Physiology **(text page 730)**

Contents

Contraceptive techniques act by blocking sperm transport, ovulation, or implantation. p. 757

Section Synopsis

Female reproductive physiology is more complex than that in males because females perform the two distinct reproductive functions of gamete production and providing an environment for fertilization and subsequent maintenance of the embryo until it can survive in the outside world. Regulation of these two functions is entirely under hormonal control. In the nonpregnant state, reproductive function is controlled by a complex, cyclical negative-feedback control system between the hypothalamus (GnRH), anterior pituitary (FSH and LH), and ovaries (estrogen, progesterone, and inhibin). During pregnancy, placental hormones become the main controlling factors.

The ovaries perform the dual and interrelated functions of producing ova (oogenesis) and secreting estrogen and progesterone. Two related ovarian endocrine units sequentially accomplish these functions: the follicle and the corpus luteum. A female is born with a non-replenishable reservoir of ovarian follicles, each containing a primary oocyte arrested in the first stage of meiosis. Oogenesis and estrogen secretion take place within an ovarian follicle during the first half of each reproductive cycle. At approximately midcycle, the maturing follicle releases a single ovum (ovulation). The empty follicle is then converted into a corpus luteum, which produces progesterone as well as estrogen during the last half of the cycle. This endocrine unit is responsible for preparing the uterus as a suitable site for implantation should the released ovum be fertilized. If fertilization and implantation do not occur, the corpus luteum degenerates. The consequent withdrawal of hormonal support for the highly developed uterine lining causes it to disintegrate and slough, producing menstrual flow. Simultaneously, a new wave of follicular development is initiated under the influence of FSH and LH as ovarian steroid inhibition of these anterior pituitary hormones is withdrawn. Menstruation ceases and the uterine lining (endometrium) repairs itself under the influence of rising estrogen levels from the newly maturing follicle.

If fertilization does take place, it occurs in the oviduct as the released egg and sperm deposited in the vagina are both transported to this site. At the time of ovulation, the egg is in the secondary oocyte stage. Just before ovulation, the first meiotic division is completed and a secondary oocyte containing a half set of doubled chromosomes is released. A polar body containing the other half of the chromosomes but no cytoplasm is produced and disintegrates. Fertilization triggers the second meiotic division, resulting in an ovum with a half set of unpaired chromosomes. A second polar body containing the remaining unused chromosomes is extruded. The twenty-three maternal chromosomes in the ovum unite with the twenty-three father-furnished chromosomes in the first spermatozoon to penetrate the protective layers surrounding the egg to complete fertilization. The formation of a sperm-proof membrane around the fertilized ovum blocks further sperm entry.

The fertilized ovum begins to divide mitotically. Within a week it grows and differentiates into a blastocyst capable of implantation. Meanwhile, the endometrium has become richly vascularized and stocked with stored glycogen under the influence of luteal-phase progesterone. It is into this especially prepared lining that the blastocyst implants. The blastocyst is a sphere containing two different cell types surrounding a fluid-filled cavity. The dense inner cell mass located at one side of the blastocyst will become the fetus. The fluid-filled cavity will become the amniotic sac within which the fetus floats. The outer trophoblastic layer releases enzymes that digest nutrient-rich endometrial tissue in advance of penetrating fingerlike cords of trophoblastic cells. This action accomplishes the dual function of carving out a hole in the endometrium for implantation of the blastocyst while at the same time releasing nutrients from the endometrial cells for use by the developing embryo.

Following implantation, the trophoblast continues to expand into the endometrium, eroding endometrial blood vessels that ooze blood. Blood vessels linked to the fetal circulation by way of the umbilical cord extend into these trophoblastic projections to bring fetal blood into close contact with maternal blood. This interlocking combination of fetal and maternal tissues constitutes the placenta, the organ of exchange between the maternal and fetal blood.

The placenta also acts as a transient, complex endocrine organ that secretes a number of hormones essential for pregnancy, the most important of which are chorionic gonadotropin, estrogen, and progesterone. Chorionic gonadotropin maintains the corpus luteum of pregnancy, stimulating its secretion of estrogen and progesterone during the first ten weeks of pregnancy until the placenta is able to take over production of these two steroids.

At parturition, rhythmic contractions of increasing strength, duration, and frequency accomplish the three stages of labor: dilation of the cervix, birth of the baby, and delivery of the placenta (afterbirth). The trigger for the onset of labor is uncertain but is believed to be an estrogen-induced increase in oxytocin receptors in the uterine musculature (myometrium) to a critical threshold level that renders the uterus responsive to the normal circulating levels of oxytocin. Once the contractions are initiated, a positive-feedback cycle is established that progressively increases their force. As contractions push the fetus against the cervix, secretion of oxytocin, a powerful myometrial stimulant, is reflexly increased. The extra oxytocin causes stronger contractions, giving rise to even more oxytocin release, and so on. Oxytocin also triggers local production of a prostaglandin that further stimulates uterine contractions. This positive-feedback cycle progressively intensifies until cervical dilation and delivery are accomplished.

During gestation, the breasts are specially prepared for lactation. The elevated levels of placental estrogen and progesterone respectively promote development of the ducts and alveoli in the mammary glands. Prolactin stimulates the synthesis of enzymes essential for milk production by the alveolar epithelial cells. However, the high gestational level of estrogen and progesterone prevents prolactin from promoting milk production. Withdrawal of the placental steroids at parturition initiates lactation. Lactation is sustained by suckling, which triggers the release of oxytocin and prolactin. Oxytocin causes milk ejection by stimulating the myoepithelial cells surrounding the alveoli to squeeze the secreted milk out through the ducts. Prolactin stimulates the production of more milk to replace the milk ejected as the baby nurses. Thus, the female reproductive system supports the new being from the moment of its conception through nourishing it during its early life outside of the supportive uterine environment.

Learning Check (Answers on p. A-66)

A. True/False

T/F 1. One primary oocyte containing 46 doubled chromosomes yields only one ovum containing 23 unpaired chromosomes.

T/F 2. If a follicle fails to reach maturity during one ovarian cycle, it can finish maturing during the next cycle.

T/F 3. Low, rising levels of estrogen inhibit tonic LH secretion whereas high levels of estrogen stimulate the LH surge.

T/F 4. The vagina is the site of fertilization.

T/F 5. Sperm are able to penetrate the zona pellucida only after binding with specific receptor sites on the surface of this layer.

T/F 6. The third stage of labor involves the actual birth of the baby.

B. Fill-in-the-blank

1. A surge in ________________ secretion from the anterior pituitary triggers ovulation.

2. Following ovulation, the ruptured follicle is transformed into the ________________.

3. During estrogen production by the follicle, the ___________ cells under the influence of the hormone _______________ produce androgens and the ________________ cells under the influence of the hormone _________________ convert these androgens into estrogens.

4. During gestation, the __________________ produces dehydroepiandrosterone, which the ______________________ converts into estrogen.

5. The __________________ of the blastocyst is destined to become the embryo/fetus.

6. The source of estrogen and progesterone during the first ten weeks of gestation is the ____________________.

7. Detection of _______________ in the urine is the basis of pregnancy diagnosis tests.

C. Indicate when the event in question takes place during the ovarian cycle using the answer code below:

a = occurs during the follicular phase

b = occurs during the luteal phase

c = occurs during both the follicular and luteal phases

___ 1. development of antral follicles

___ 2. secretion of estrogen

___ 3. secretion of progesterone

___ 4. menstruation

___ 5. repair and proliferation of the endometrium

___ 6. increased vascularization and glycogen storage in the endometrium

D. Indicate whether the following functions are attributable to estrogen or progesterone using the answer code below:

a = estrogen

b = progesterone

c = both estrogen and progesterone

___ 1. responsible for developing and maintaining female secondary sex characteristics

___ 2. stimulates duct development in the breasts

___ 3. inhibits uterine contractility

___ 4. causes cervical mucus to become thick and sticky

___ 5. induces endometrial secretory capacity

___ 6. promotes thickening of the myometrium

___ 7. promotes antiperistaltic contractions in the oviduct

___ 8. inhibits prolactin's ability to promote milk secretion

E. Multiple choice

1. Which of the following statements concerning oogenesis is (are) correct?

   a. The first meiotic division occurs just prior to ovulation and the second meiotic division is triggered by fertilization.

   b. The process of oogenesis takes anywhere from 12 to 50 years to complete.

   c. Oogonia proliferate mitotically throughout the reproductive life of a female.

   d. Both a and b above are correct.

   e. All of the above are correct.

2. When the corpus luteum degenerates

   a. circulating levels of estrogen and progesterone rapidly decline

   b. FSH and LH secretion start to rise as the inhibitory effects of the gonadal steroids are withdrawn

   c. the endometrium sloughs

   d. Both a and b are correct.

   e. All of the above are correct.

3. Which of the following statements concerning implantation is (are) correct?

   a. Implantation is accomplished by enzymatic activity of the trophoblastic layer of the blastocyst.

   b. The endometrium at the site of implantation is converted into the nutrient-rich decidua.

   c. Implantation occurs within 24 hours after fertilization.

   d. Both a and b above are correct.

   e. All of the above are correct.

4. Which of the following statements concerning the placenta is (are) correct?

   a. The placenta secretes human chorionic gonadotropin, estrogen, and progesterone.

   b. The placenta serves the functions of the circulatory, respiratory, and digestive systems for the fetus.

   c. Maternal and fetal blood are mixed together within the placenta.

   d. Both a and b above are correct.

   e. All of the above are correct.

5. Oxytocin

   a. is a powerful uterine muscle stimulant

   b. is involved in a positive feedback cycle during parturition

   c. stimulates production of milk by the mammary glands

   d. Both a and b above are correct.

   e. All of the above are correct

Chapter in Perspective (text page 759)

APPENDIX A: ANSWERS TO LEARNING CHECKS

Chapter 1 Homeostasis: The Foundation of Physiology

Levels of Organization in the Body (manual page 2)

A. 1.e, 2.b, 3.c

B. 1. F (A teleological explanation of why a person sweats is to cool off. A mechanistic explanation is that when temperature-sensitive nerve cells signal the hypothalamus that there has been a rise in body temperature, the hypothalamus stimulates the sweat glands to produce sweat, evaporation of which cools the body.)

2. T

3. T

4. F (Unicellular organisms and cells in a multicellular organism all perform similar basic functions essential for the cell's survival.)

5. T

C. 1. cell

2. muscle tissue, nervous tissue, epithelial tissue, connective tissue

3. secretion

4. exocrine, endocrine, hormones

Concept of Homeostasis (manual page 5)

A. 1. internal environment

2. extracellular fluid, plasma, interstitial fluid

3. intrinsic, extrinsic

4. negative feedback

5. pathophysiology

B. concentration of nutrient molecules, concentration of O_2 and CO_2, concentration of waste products, pH, concentration of salt and other electrolytes, temperature, volume and pressure

C. 1. T

2. T

3. F (The internal environment must be maintained in a dynamic steady state instead of a fixed state. Fluctuations around an optimal level are minimized by compensatory physiological responses so that the homeostatically maintained factors are kept within the narrow limits that are compatible with life.)

D. 1.d, 2.g, 3.a, 4.e, 5.b, 6.j, 7.h, 8.i, 9.c, 10.f

Chapter 2 Cellular Structure and Functions

Introduction (manual page 8)

A. 1. T

2. F (Human cells cannot be seen by the unaided eye; they are about ten times smaller than the smallest point visible to the naked eye.)

3. F (Plant cells, not human or other animal cells, are surrounded by a cell wall.)

B. 1. plasma membrane

2. deoxyribonucleic acid (DNA), nucleus

3. organelles, cytosol, cytoskeleton

Organelles (manual page 11)

A. 1. T

2. F (Coated vesicles capture specific cargo, not a representative mixture of proteins, before budding off from the Golgi sac.)

3. T

4. F (Lysosomal enzymes are less potent in the cytosol than in the acidic environment within the lysosome. Furthermore, usually cells can tolerate the limited damage accompanying the inadvertent rupture of a few lysosomes because most cell parts are renewable.)

5. T

B. 1. proteins, lipids

2. leader sequence, ribophorin

3. smooth

4. endoplasmic reticulum, Golgi complex

5. coated vesicles, secretory vesicles

6. exocytosis

7. hydrolytic

8. endocytosis

9. oxidative, hydrogen peroxide

10. catalase

11. adenosine triphosphate (ATP)

C. 1.b, 2.a, 3.b

D. 1.b, 2.c, 3.c, 4.a, 5.b, 6.c, 7.a, 8.c

Cytosol and Cytoskeleton (manual page 14)

A. 3

B. 1.b, 2.c, 3.b, 4.c, 5.a, 6.a, 7.b

C. 1.a, 2.c, 3.b, 4.b, 5.d, 6.a, 7.a, 8.a, 9.a, 10.d, 11.b, 12.a

D. 1. T

2. T

3. F (The protective, waterproof outer layer of skin is formed by the persistence of tough keratin, not microtrabecular lattice, skeletons when the surface skin cells die. Keratin is a protein component of special intermediate filaments found in skin cells.)

Nucleus (manual page 18)

A. 1. deoxyribonucleic acid (DNA)

2. thymine, adenine, guanine, cytosine

3. gene

4. 46, 23

5. histone

6. chromatin

7. triplet code

8. initiation, elongation, termination

9. mitosis, meiosis

10. sister chromatids

11. cytokinesis (cytoplasmic division)

12. interphase

13. crossing over

14. hormones

B. 1. T

2. F (Many ribosomes can attach simultaneously to a single mRNA molecule, with each ribosome, one behind the other, synthesizing the same type of protein as specified by the mRNA code.)

3. F (Mutations can be harmless, deleterious, fatal, or beneficial.)

C. 1.b, 2.c, 3.a, 4.d

D. 1.b, 2.c, 3.a, 4.b, 5.a, 6.b, 7.c

E. 1.a, 2.c

Chapter 3 Plasma Membrane and Membrane Potential

Membrane Structure and Composition (manual page 24)

A. 1. T

2. F (The hydrophobic regions correspond to the light space; the hydrophilic regions correspond to the two dark layers.)

3. F (Membrane components are arranged asymmetrically.)

B. 1.b, 2.a, 3.b, 4.a, 5.c, 6.b, 7.a, 8.b

Membrane Receptors and Postreceptor Events (manual page 26)

A. 1. channel regulation mechanism, second messenger mechanism

2. cyclic AMP, Ca^{++}

B. 1.b, 2.c, 3.a, 4.d

C. 1. T

2. F (A tremendous amplification of the initial signal occurs as a result of a cascade effect.)

Cell-to-Cell Adhesions (manual page 27)

A. 1.a, 2.d, 3.c, 4.b

B. 1.c, 2.b, 3.a, 4.a, 5.c, 6.b, 7.c, 8.a, 9.b

Membrane Transport (manual page 29)

A. 1. F (There is no net movement; at equilibrium an equal number of molecules are moving in opposite directions across the membrane.)

2. T

B. 1.a, 2.a, 3.b, 4.a, 5.b, 6.a, 7.b

C. 1. pinocytosis, phagocytosis, endocytosis

2. Ca^{++}

Membrane Potential (manual page 32)

A. 1. membrane potential

2. concentration

3. negative, positive

B. 1. F (A potential of 70 mV is of greater magnitude than a potential of 30 mV. The sign refers to the charge on the inside of the cell.)

2. F (Only 20% of the membrane potential is <u>directly</u> generated by the Na^+-K^+ pump. The other 80% is caused by the passive diffusion of K^+ and Na^+ down concentration gradients. The Na^+-K^+ pump <u>indirectly</u> contributes to membrane potential by maintaining these concentration gradients.)

3. T

C. 1.b, 2.a, 3.c, 4.d, 5.b, 6.a, 7.d, 8.b, 9.c, 10.b, 11.a,

12.b, 13.d

Chapter 4 Neuronal Physiology

Electrical Signals: Graded Potentials and Action Potentials (manual page 37)

A. 1.b, 2.a, 3.a, 4.b, 5.b, 6.a

B. 1. T

2. F (The membrane is returned to resting potential after it reaches the peak of an action potential as a result of K^+ efflux. The Na^+-K^+ pump gradually restores the original concentration gradients following an action potential by pumping out the Na^+ and pumping in the K^+ that left during the action potential.)

3. F (Following an action potential, there is still much more K^+ inside than outside the cell despite the fact that K^+ leaves during the action potential because only an extremely small percentage of the total intracellular K^+ leaves.)

C. 1.a; 2.b,c; 3.e; 4.h

D. 1. conduction by local current flow; * saltatory conduction

2. refractory period

Synapses and Neuronal Integration (manual page 41)

A. 1. synapse

2. Ca^{++}

3. temporal summation

4. spatial summation

5. axon hillock

6. convergence, divergence

B. 1.a, 2.b, 3.a, 4.b

C. 1. F (Some neurotransmitters function through intracellular second messengers rather than by directly altering membrane permeability.)

2. F (Synapses operate in one direction only; the presynaptic neuron influences the postsynaptic neuron, but the postsynaptic neuron cannot influence the presynaptic neuron.)

3. F (A given synapse is either always excitatory (i.e, can produce only EPSPs) or always inhibitory (i.e., can produce only IPSPs).)

4. T

Chapter 5 Central Nervous System

Introduction (manual page 44)

A. nervous, endocrine

B. 1.d, 2.c, 3.f, 4.e, 5.a, 6.b

C. 1.a; 2.c; 3.a,b; 4.b; 5.a; 6.c; 7.c

Protection and Nourishment of the Brain (manual page 46)

A. 1.d, 2.c, 3.b, 4.a

B. 1.e, 2.d, 3.a, 4.c, 5.b

C. 1. F (The major function of CSF is to serve as a shock-absorbing fluid to prevent brain damage during sudden, jarring movements of the head.)

2. F (The brain cannot produce ATP under anaerobic conditions.)

3. T

Cerebral Cortex (manual page 49)

A. 1. T

2. F (Damage to the left hemisphere brings about paralysis and loss of sensation on the right side of the body because of fiber cross-over.)

3. T

4. F (The right hemisphere specializes in verbal and analytical skills whereas the left side excels in artistic and musical ability.)

5. F (The brain has considerable plasticity; the organizational pattern of sensory and motor regions of the cortex is subject to constant subtle modifications based on use-dependent competition for cortical space.)

6. F (The EEG is a record of all postsynaptic activity in the cortex, most of which is EPSPs and IPSPs, not action potentials.)

. 1.e, 2.g, 3.a, 4.c, 5.i, 6.f, 7.h, 8.b, 9.d, 10.j

ubcortical Structures and Their Relationship with the Cortex in Higher Brain Functions (manual page 52)

. 1. motivation

2. learning

3. consolidation

4. memory trace or engram

5. retrograde amnesia, anterograde amnesia

6. habituation

7. sensitization

. 1.c, 2.e, 3.a, 4.b, 5.d

. 1.b, 2.a, 3.a, 4.b, 5.a, 6.b

erebellum; Brain Stem (manual page 54)

. 1. intention; 2. consciousness

. 1.b, 2.c, 3.a

. 1.b, 2.a, 3.a, 4.b, 5.b, 6.a, 7.b, 8.a, 9.a, 10.b, 11.b, 12.b

pinal Cord (manual page 57)

. 1.b, 2.a, 3.c, 4.a, 5.b, 6.a

. 1. T; 2. T

. 1. dorsal, ventral

2. receptor, afferent pathway, integrating center, efferent pathway, effector

Chapter 6 Peripheral Nervous System: Afferent Division; Special Senses

Introduction (manual page 60)

A. 1. transduction; 2. adequate stimulus

B. 1. T

2. F (Humans only have receptors to detect a limited number of existing energy forms.)

3. F (Only afferent information that reaches conscious levels of the brain is considered to be sensory information.)

C. 1. Distinguished by the type of receptor activated and the specific pathway (i.e., labeled line) over which this information is transmitted to a particular area of the cerebral cortex

2. Distinguished by the location of the activated receptor field and the pathway that is subsequently activated to transmit this information to the area of the somatosensory cortex representing that particular location; enhanced by lateral inhibition

3. Distinguished by the frequency of action potentials initiated in an activated afferent neuron (frequency code) and the number of afferent neurons activated (population code)

D. 1.c, 2.d, 3.a, 4.b

Pain (manual page 62)

A. 1.a; 2.c; 3.b; 4.a,b; 5.c; 6.c; 7.a,b

B. 1. substance P, 2. opiate

C. 1.b, 2.a, 3.c

D. 1. F (Prostaglandins increase the sensitivity of nociceptors.)

2. T

Eye: Vision (manual page 65)

A. 1. T

2. F (Short wavelengths of light are perceived in the violet-blue color range; long wavelengths are perceived in the red-orange range.)

3. F (Photoreceptors and bipolar cells display graded potentials; ganglion cells are the only retinal cells that display action potentials.)

4. T

5. F (Retinal processing, such as lateral inhibition, modifies the visual input to enhance contrast.)

6. T

7. T

8. T

B. 1.f, 2.h, 3.l, 4.d, 5.i, 6.e, 7.b, 8.j, 9.a, 19.g, 11.c, 12.k

C. 1. contracts, slack, round up, increases

2. absorbs, dissociate, close, hyperpolarizes, decrease

D. 1.b; 2.a; 3.b; 4.a; 5.b; 6.a; 7.a,b; 8.a; 9.b; 10.b

E. 1.g, 2.d, 3.i, 4.g, 5.d, 6.i, 7.e, 8.c, 9.h, 10.f, 11.b, 12.a

Ear: Hearing and Equilibrium (manual page 70)

A. 1.a,b; 2.c,d,e; 3.c,d,e; 4.a,b,c; 5.d,e; 6.a; 7.b; 8.d; 9.e;

10.d; 11.e; 12.c; 13.b; 14.d,e; 15.d,e; 16.c; 17.d

B. 1.a, 2.c, 3.b, 4.c, 5.a, 6.b, 7.a, 8.b

C. 1. T

2. F (Because decibels are a logarithmic measure of sound intensity, every ten decibel increase represents a tenfold increase in loudness. Thus, a 100 dB sound is 10 billion times louder than hearing threshold.)

3. F (Mechanical deformation of the hairs in opposite directions as a result of basilar membrane oscillation triggered by sound waves results in alternating depolarization and hyperpolarization of the hair cells.)

4. F (Displacement of the round window just dissipates pressure. Displacement of the basilar membrane and the resultant activation of the organ of Corti's hair cells generates neural impulses that are perceived as sound sensations.)

5. T

6. T

7. T

8. T

9. T

Chemical Senses: Taste and Smell (manual page 74)

A. 1.d, 2.b, 3.a, 4.c

B. 1.a, 2.b, 3.c, 4.c, 5.c

C. 1. F (Each taste receptor responds in varying degrees to all four primary tastes.)

2. T

3. F (Because the olfactory mucosa is above the normal path of air flow in the nose, during quiet breathing, odoriferous molecules reach the smell receptors only by diffusion.)

4. F (Molecules of similar odor are believed to share a particular configuration, not a similar chemical conformation.)

5. F (Adaptation to odors involves some sort of CNS adaptation, not olfactory receptor adaptation.)

Chapter 7 Peripheral Nervous System: Efferent Division

Autonomic Nervous System; Somatic Nervous System (manual page 79)

A. 1.a, 2.b, 3.a, 4.b, 5.a, 6.a, 7.b

D. 1.b, 2.b, 3.a, 4.a, 5.b, 6.b, 7.a

C. 1.b, 2.a, 3.a, 4.b

D. alpha motor neuron

E. 1. F (Atropine blocks muscarinic, not nicotinic, receptors.)

2. T

Neuromuscular Junction (manual page 81)

A. 1.c, 2.b, 3.d, 4.e, 5.a

B. 3 a., 6 b., 8 c., 1 d., 5 e., 2 f., 7 g., 4 h.

Chapter 8 Muscle Physiology

Structure of Skeletal Muscle (manual page 84)

A. 1.b, 2.c, 3.d, 4.c, 5.b, 6.a, 7.d, 8.a

B. 1.a; 2.c; 3.a; 4.b; 5.d; 6. b,c,d; 7.d; 8.a,b; 9.a,b; 10. c,d

C. sarcomere

Molecular Basis of Skeletal-Muscle Contraction (manual page 87)

A. 1.a, 2.a, 3.a, 4.b, 5.b, 6.b

B. The thin filaments slide inward toward the A band's center. As they slide inward, the thin filaments pull the Z lines to which they are attached closer together, so the sarcomere shortens.

C. 1.f, 2.d, 3.c, 4.e, 5.b, 6.g, 7.a

D. b

E. F (The contractile response lasts about one-hundred times longer than the action potential.)

Gradation of Skeletal-Muscle Contraction (manual page 90)

A. 1.c, 2.b, 3.c

B. twitch

C. F (The fewer the muscle fibers within a muscle's motor units, the more precisely controlled the gradation of the muscle's contractions.)

D. 1.d, 2.a, 3.e, 4.b, 5.f, 6.c

Metabolism and Types of Skeletal-Muscle Fibers (manual page 93)

A. 1. ATP

2. creatine phosphate

3. denervation atrophy; disuse atrophy

B. 1.b,c; 2.a; 3.c; 4.a,b; 5.b,c; 6.a,b; 7.a; 8.c; 9.c; 10. c; 11.c; 12.c; 13.c; 14.c; 15.a,b; 16.a,b

Muscle Mechanics (manual page 96)

A. 1.b, 2.a, 3.b, 4.a, 5.b, 6.a, 7.b, 8.a

B. concentric, eccentric

C. F (The velocity of shortening depends on the magnitude of the load as well as on the ATPase activity of its fibers.)

Control of Motor Movement (manual page 98)

A. 1. a,b,e

2. c

B. 1. corticospinal system

2. multineuronal system

3. alpha, gamma

C. 1.a, 2.b, 3.c, 4.a, 5.a, 6.c, 7.c

Smooth and Cardiac Muscle (manual page 101)

A. 1.a,d; 2. b,c,d; 3.b,c; 4.a,d; 5.a,b; 6.c,d; 7.c,d; 8.a,d; 9.b,c,d

B. 1. T

2. F (Slow wave potentials initiate action potentials only when the automatic depolarizing swing in potential reaches threshold. Whether or not threshold is reached depends on the stating point of the membrane potential at the onset of its depolarizing swing.)

3. T

4. T

5. F (Unlike skeletal muscle, not all cross bridges are switched on by a single excitation in smooth muscle.)

6. T

7. F (Many single-unit smooth muscle cells have sufficient levels of cytosolic Ca^{++} to maintain a low level of tension, or tone, even in the absence of action potentials.)

8. T

hapter 9 Cardiac Physiology

natomical Considerations (manual page 104)

. 1.e, 2.a, 3.d, 4.b, 5.f, 6.c, 7.g

. 1. F (The left ventricle is a stronger pump because it pumps the same quantity of blood as the right ventricle does but must pump into the high-pressure, high-resistance systemic circulation whereas the right ventricle pumps into the low-pressure, low-resistance pulmonary circulation.)

2. F (The heart is located approximately midline within the thoracic cavity.)

3. T

4. T

. 1. intercalated discs, desmosomes, gap junctions

2. cardiac tamponade

lectrical Activity of the Heart (manual page 107)

. 1. contraction, initiating and conducting the action potential

2. a cyclical decrease in the passive outward flux of K^+ resulting from inactivation of K^+ channels between action potentials

3. SA node

4. AV-nodal delay

5. the long refractory period

6. bradycardia, tachycardia

B. 1.c, 2.a, 3.b

C. 1. F (The only point of electrical contact between the atria and ventricles is the AV node.)

2. T

3. F (The ECG is a recording of that portion of the electrical activity induced in the body fluids by the cardiac impulse that reaches the surface of the body, not a direct recording of the actual electrical activity of the heart.)

4. T

5. T

D. 4

E. 1.b, 2.f, 3.a, 4.b, 5.d, 6.e

Mechanical Events of the Cardiac Cycle (manual page 110)

A. 1. less than, greater than, less than, greater than, less than

2. insufficient, semilunar

3. AV, systole, semilunar, diastole

Cardiac Output and Its Control (manual page 112)

A. 1.a, 2.a, 3.a, 4.b, 5.a, 6.b, 7.b, 8.a, 9.a, 10.a, 11.a,

12.b, 13.b, 14.a, 15.c, 16.a, 17.a, 18.a, 19.b, 20.a

Nourishing the Heart Muscle (manual page 115)

A. 1. F (Most coronary blood flow occurs during ventricular diastole; during systole, the coronary vessels are compressed almost closed by the contracting heart muscle.)

2. F (The heart primarily uses free fatty acids but can also use glucose or lactate, depending on their availability, for energy production.)

3. T

B. 1. adenosine

2. ischemia

3. the size of the occluded vessel, the extent of collateral circulation

4. immediate death, delayed death from complications, full functional recovery, recovery with impaired function

5. LDL, HDL

6. liver, bile salts

7. blood HDL-cholesterol/total cholesterol ratio

C. 1.c, 2.a, 3.b, 4.d

Chapter 10 The Vasculature and Blood Pressure

Introduction (manual page 118)

A. 1. T

2. T

3. T

B. 1.a, 2.b, 3.a, 4.b, 5.b

Arteries (manual page 120)

A. passageway to tissues, pressure reservoir

B. 1. F (Blood flow is continuous through the capillaries. The driving force for continued blood flow to the tissues during cardiac diastole is provided by the elastic recoil of the arterial walls that had been stretched by blood during cardiac systole.)

2. T

C. 1. 125, 2. 77, 3. 48, 4. 93, 5. no, 6. yes, 7. no

Arterioles (manual page 122)

A. 1. a,c,d,e,f

2. b,e,f

B. 1. tone

2. increase, vasodilation, decrease, vasoconstriction

3. local

4. vasopressin, angiotensin II

C. 1.b, 2.b, 3.b, 4.a, 5.b, 6.c, 7.a, 8.c, 9.a

Capillaries (manual page 126)

A. 1. F (Whether or not blood is flowing through a given capillary bed depends on the status of the region's arterioles and precapillary sphincters.)

2. T

3. T

4. T

5. F (Capillaries vary in pore size.)

6. T

B. 1. passive diffusion down concentration gradients

2. bulk flow

3. plasma proteins

4. colloidal suspension of plasma proteins that remain trapped within the plasma

5. picked up by the lymphatic system and returned to the circulatory system

C. 1. ultrafiltration

2. reabsorption

3. edema

D. 1.a, 2.a, 3.b, 4.a, 5.a, 6.a

Veins (manual page 129)

A. 1, 3, 5, 6, 7, 9, 10, 12

B. 1.a, 2.a, 3.b, 4.a, 5.b, 6.a

C. 1. F (Sufficient driving pressure imparted to the blood by cardiac contraction is still present in the veins to continue driving blood forward.)

2. T

3. T

4. F (Varicose veins have little effect on effective circulating volume.

5. F (Assumption of a horizontal position upon fainting removes the gravitational effects on the vasculature and restores an effective circulating blood volume. Thus, holding a fainted person upright is counterproductive.)

Blood Pressure (manual page 132)

A. 1. cardiac output, total peripheral resistance

2. heart rate, stroke volume

3. the degree of arteriolar vasoconstriction

4. carotid sinus; aortic arch; cardiovascular control center; medulla within the brain stem; autonomic nervous system; heart, blood vessels (arterioles and veins)

B. 1.b, 2.a, 3.b, 4.a, 5.a, 6.a, 7.b, 8.a, 9.b, 10.a, 11.b, 12.a, 13.a

C. 1. F (Evidence suggests that discrete exercise centers induce the appropriate cardiovascular changes at the onset of exercise.)

2. T

3. F (In the presence of chronically elevated blood pressure, the baroreceptors are "reset" to maintain the blood pressure at the higher mean pressure.)

4. T

5. T

D. 1. a,c,d,f,g

2. a,b,c,d,e,f,g

Chapter 11 Blood

Plasma (manual page 136)

A. 1. hematocrit, less than

2. erythrocytes (red blood cells), leukocytes (white blood cells), thrombocytes (platelets)

3. erythrocytes

B. 1. T

2. F (Plasma proteins are not freely diffusible across the capillary walls; they are too large to pass through the capillaries' pores.)

C. 4

Erythrocytes (manual page 138)

A. 1. F (Hemoglobin normally carries CO_2 and H^+ from the tissues to the lungs. It also has a high affinity for carbon monoxide.)

B. 1. carbonic anhydrase

2. spleen

3. bone marrow, erythropoietin, reduced O_2 delivery

C. 1.e, 2.c, 3.b, 4.d, 5.g, 6.f, 7.a, 8.h

D. 1. b

2. b

Leukocytes (manual page 141)

A. 1. by phagocytosis; by immune responses, such as the production of antibodies that mark invaders for destruction in more subtle ways.)

2. lymphocytes

3. neutrophils, eosinophils, basophils

B. 1. F (White blood cells are in the blood only while in transit from their site of production and storage in the bone marrow or lymphoid organs to their site of action in the tissues.)

2. T

C. 1.a, 2.d, 3.e, 4.a, 5.c, 6.b, 7.e, 8.e, 9.c

Platelets and Hemostasis (manual page 144)

A. 1. F (Platelets lack nuclei but they do have organelles and cytosolic enzyme systems for generating energy.)

2. T

3. F (Plasma contains the unactivated clotting factors but serum lacks them.)

B. 1. vascular spasm, formation of a platelet plug, blood coagulation

2. liver

C. a

D. 1.d, 2.g, 3.c, 4.a, 5.h, 6.e, 7.i, 8.f, 9.g, 10.b

Chapter 12 Defense Mechanisms of the Body

Introduction (manual page 148)

A. 1. b

2. b

B. 1. b

2. a

3. b

4. a

5. b

6. b

C. 1. F (The spleen clears the blood, not the lymph, that passes through it of bacteria and other foreign matter.)

2. T

3. F (The complement system can also be activated by the presence of any foreign invader.)

4. F (Specific immune responses are accomplished by lymphocytes.)

D. 1. c

2. d

3. a

4. b

Nonspecific Immune Responses (manual page 152)

A. 1. a, e

2. d

3. c, e

4. b

B. 1. T

2. F (Phagocytes can also destroy invaders by several nonphagocytic, extracellular means, such as by releasing destructive chemicals.)

3. T

4. F (In tissues capable of regeneration, lost cells are replaced by cell division of surrounding, healthy, organ-specific cells so that perfect repair is accomplished.)

5. F (Interferon is released from any virus-invaded cell.)

6. T

C. 1. margination

2. diapedesis

3. chemotaxis

4. granuloma

5. pus

6. opsonin

7. natural killer (NK) cells

8. membrane-attack complex

9. inflammation

D. 1. salicylates, glucocorticoids

2. -extracellular release of destructive lysosomal enzymes;

-release of lactoferrin, which binds iron needed for bacterial multiplication;

-lysing of bacteria by the complement system's membrane-attack complex

3. -classical pathway; antibodies activate complement system

-alternate pathway; particular carbohydrate chains on microbial surfaces activate complement system

Specific Immune Responses (manual page 159)

A. 1. a

2. a

3. b

4. b

5. c

6. c

7. b

8. a

9. b

10. b

11. a

12. b

B. 1. a

2. b

3. a

4. b

C. 1. b

2. d

3. a

4. c

5. b

D. 1. F (Active immunity can also be acquired through vaccination.)

2. T

3. F (Type AB blood can be donated only to Type AB individuals because its erythrocytes contain both A and B antigens that will react with the antibodies found in all other ABO blood types. Because Type AB blood lacks both anti-A and anti-B antibodies, individuals of this type can receive blood of any other ABO type.)

4. T

5. T

6. F (Several sequential mutations are required to convert a normal cell into a cancer cell.)

E. 1. antigen

2. hapten

3. plasma cells

4. agglutination

5. clonal selection

6. macrophages

7. helper T cell

8. major histocompatibility complex (MHC)

9. human leukocyte-associated (HLA) antigens

10. lymphokines

11. passive

12. benign, malignant, cancer, metastasis

Immune Diseases (manual page 164)

A. 1. F (Slow-reactive substance of anaphylaxis (SRS-A) is primarily responsible for causing the bronchial constriction of asthma.)

2. F (Eosinophils are attracted to sites involved with immediate hypersensitivity reactions.)

3. T

B. 1. immune complex

2. IgE

3. Anaphylactic shock

External Defenses (manual page 167)

A. 1. b

2. a

3. a

4. b

5. a

6. a

7. a

8. b

B. 1. T

2. T

3. F (Saliva is destructive to bacteria because it contains lysozyme, an enzyme that lyses certain bacteria.)

4. T

5. F (Debris-laden mucus is swept out of the respiratory airways by ciliary action; that is, by the mucus escalator.)

6. F (A sneeze expels irritant material from the nose.)

C. 1. b

2. c

3. a

4. a

5. b

6. c

7. d

8. a

9. a

D. 1. a. supply blood to both the dermis and epidermis

b. play a major role in temperature regulation; the volume of blood flowing through these vessels can be adjusted to vary the amount of heat exchange between the skin and external environment.

2. a. sweat; important in temperature regulation by cooling the skin

b. sebum; oils hairs and outer layers of skin, which waterproofs and softens them

c. hairs; increase the skin's sensitivity to touch and important in lower species in heat conservation

Chapter 13 Respiratory System

Introduction (manual page 170)

A. 1. respiratory airways, alveoli

2. CO_2 produced, O_2 consumed

3. external respiration

4. pharynx

5. one, one

6. collateral ventilation

7. diaphragm

8. pleural sac

B. 1. F (Both the respiratory and circulatory systems are essential for accomplishing external respiration.)

2. F (The respiratory muscles directly change the size of the thoracic cavity, not the lungs. There are no muscles in the lungs, except for the smooth muscle in the walls of the airways and blood vessels.)

3. T

4. F (Unrelated to their role in speech, the vocal cords close off the entrance to the trachea during swallowing.)

5. F (The smaller bronchioles have no cartilage to hold them open.)

6. T

Respiratory Mechanics **(manual page 175)**

A. 1. transmural pressure gradient, pulmonary surfactant, alveolar interdependence

2. pulmonary elasticity, alveolar surface tension

3. emphysema

4. compliance

5. elastic recoil

6. vital capacity

7. anatomic dead space

B. 1. F (A volume of air, known as the residual volume, remains in the lungs after a maximal expiration because air is trapped in the alveoli as a result of compression of the small nonrigid airways by the high intrapleural pressure accompanying the forceful expiration.)

2. T

3. T

4. F (An increase in pulmonary ventilation may be associated with an increase in movement of air in and out of the anatomic dead space, with no increase in alveolar ventilation.)

5. F (Only 350 ml of fresh inspired air out of a tidal volume of 500 ml enters the alveoli because of the volume of air that occupies the anatomic dead space.)

C. 1. constricts

2. dilates

3. dilates

4. dilates

D. 1.>, 2.<, 3.<, 4.=, 5.<, 6.>, 7.=, 8.<, 9.>, 10.<, 11.<,

12.>, 13.<, 14.=, 15.<, 16.<, 17.<, 18.>, 19.<

Gas Exchange (manual page 180)

A. 1. T

2. T

3. T

4. F (The diffusion coefficient for CO_2 is twenty times that of O_2.)

B. 1.b, 2.b, 3.b, 4.c, 5.a, 6.b, 7.a

C. 100 mm Hg

D. 1.<, 2.>, 3.=, 4.=, 5.=, 6.=, 7.>, 8.<, 9. approximately =, 10. approximately =, 11.=, 12.=

Gas Transport (manual page 184)

A. 1. a. physically dissolved, 1.5

bound to hemoglobin, 98.5

b. physically dissolved, 10

bound to hemoglobin, 30

as bicarbonate, 60

2. O_2, CO_2, H^+, CO

B. 1. Po_2 of the blood

2. carbonic anhydrase

3. chloride

4. Bohr

5. Haldane

6. hyperventilation

C. 1. T

2. T

3. F (The combination of Hb and CO_2 is known as carbamino hemoglobin. The combination of Hb and CO is known as carboxyhemoglobin.)

4. F (Hemoglobin has a much higher affinity for CO than for O_2.)

5. F (Hypercapnia does not accompany anemic hypoxia, histotoxic hypoxia, or hypoxic hypoxia due to high altitude.)

D. 1. a,c,d,e

2. a,d

E. 1.b; 2.a,f; 3.e; 4.d; 5.a; 6.e; 7.c; 8.g; 9.b; 10.d; 11.c;

12.f; 13.f; 14.g

Control of Respiration (manual page 189)

A. 1. F (Rhythmicity of breathing is brought about by pacemaker activity of the inspiratory neurons, not by the respiratory muscles themselves.)

2. F (Impulses from the expiratory neurons to the expiratory muscles occur only during active expiration, not during normal quiet breathing.)

3. T

4. T

5. T

6. T

7. F (The cerebral cortex sends impulses directly to the motor neurons in the spinal cord that supply the respiratory muscles.)

B. 1. medulla of the brain stem

2. pons of the brain stem

3. phrenic

4. inspiratory

5. VRG

6. arterial Pco_2

7. carotid, aortic

8. apnea

9. dyspnea

C. 1. b

2. d

D. 1.d, 2.a, 3.b, 4.a, 5.b, 6.a

Chapter 14 Urinary System

Introduction (manual page 195)

A. 1. T

2. T

3. F (As the kidney tubules act on the plasma to maintain constancy in ECF composition and volume, they produce urine of varying composition and volume.)

4. T

5. T

B. 1. b (The kidneys act directly on the plasma, not the interstitial fluid.)

2. a (Glomeruli of all nephrons are in the cortex.)

C. 1. nephron

2. cortex, medulla

3. 20

D. 1.b, 2.d, 3.a, 4.c

E. 1. b,e,a,d,c

2. c,e,d,a,b,f

3. c,d,a,f,b,e

Glomerular Filtration (manual page 199)

A. 1. F (Glomerular filtration is a passive process, requiring no energy expenditure by the kidneys.)

2. F (Bowman's capsule contains filtration slits between the foot processes of adjacent podocytes. The glomerular capillary wall contains pores between adjacent endothelial cells.)

3. F (The pores permit passage of albumin, but it is repelled by the charged glycoproteins in the basement membrane.)

4. T

5. F (Even though autoregulaion strives to keep the GFR constant in the arterial blood pressure range of 80 to 180 mm Hg, extrinsic control measures can deliberately alter the GFR in this range as part of the baroreceptor reflex response for blood pressure regulation.)

6. T

7. T

B. 1. K_f, net filtration pressure

2. 125

3. macula densa, juxtaglomerular apparatus, afferent, tubulo-glomerular

C. 1.b, 2.a, 3.b, 4.b, 5.a, 6.b, 7.b, 8.b, 9.b

Tubular Reabsorption and Tubular Secretion (manual page 204)

A. 1. luminal membrane of tubular cell
2. cytosol of tubular cell
3. basolateral membrane of tubular cell
4. interstitial fluid
5. capillary wall

B. 1.b, 2.d, 3.e

C. 1.a, 2.a, 3.a

D. 1.e, 2.b, 3.e, 4.b, 5.c, 6.e, 7.e, 8.a, 9.f, 10.e, 11.d,

E. 1. F (For a substance to be actively reabsorbed, any one of the five steps of transepithelial transport requires energy expenditure.)

2. F (Sodium reabsorption is not under hormonal control in the proximal tubule and loop of Henle. Aldosterone increases sodium reabsorption in the distal and collecting tubules.)

3. T

4. F (Urea is passively reabsorbed. The tubular cells display a T_m only for substances they actively reabsorb.)

5. T

6. T

7. T

8. F (The organic-ion secretory systems in the proximal tubule facilitate elimination of foreign organic compounds from the body.)

F. 1. Na^+-K^+ ATPase pump, basolateral

2. 250 mg/min

3. 200 mg/min, 50 mg/min

4. renal threshold

5. potassium

6. liver

Urine Excretion and Plasma Clearance (manual page 210)

A. 1. 124, 1

2. inulin

3. para-aminohippuric acid

4. twenty, vasopressin

5. 500

6. hypothalamic osmoreceptors, left atrial baroreceptors

7. antidiuretic hormone (ADH)

8. water diuresis, osmotic diuresis

9. acute, chronic

10. twenty-five

11. urinary incontinence

B. 1. T

2. T

3. T

4. T

5. T

6. T

7. T

8. F (The epithelial lining expands by means of insertion of membrane-bound vesicles into the lining by exocytosis.)

9. F (One can voluntarily prevent urination in spite of reflex bladder contraction by deliberate tightening of the external sphincter and pelvic diaphragm.)

C. 1. hypotonic, 2. inhibited, 3. large, 4. hypotonic

D. 1.a, 2.a, 3.c, 4.b, 5.d

E. 1.b, 2.e, 3.c

F. 1.b; 2.a; 3.a,a; 4.b; 5.a

Chapter 15 Fluid Balance and Acid-Base Balance

Introduction; Fluid Balance (manual page 217)

A. 1. T

2. F (Osmolarity is determined by the number of solute particles within a given volume of fluid, not by the mass of the solute particles.)

3. T

4. F (With an isotonic fluid gain, there is no change in ECF osmolarity to induce a fluid shift between the ECF and cells, so the fluid gain is confined to the ECF compartment.)

5. T

B. 1. 60

2. pool

3. stable

4. extracellular fluid

5. interstitial fluid, plasma

6. interstitial fluid

7. transcellular fluid

8. blood vessel walls

9. capillary blood pressure, colloid osmotic pressure

10. plasma membrane of cells

11. osmotic pressure differences

12. sodium, chloride, bicarbonate

13. potassium, phosphate, negatively charged intracellular proteins

14. the amount of Na^+ filtered, the amount of Na^+ reabsorbed

15. GFR

16. renin-angiotensin-aldosterone

17. vasopressin

C. 1.d, 2.b

D. Sources of H_2O input

X Fluid intake
H_2O in food intake
Metabolically produced H_2O

Sources of H_2O output

Insensible loss from lungs and non-sweating skin
Sweat
Feces
* X Urine

E. 1.b, 2.a, 3.a, 4.b, 5.b, 6.a

F. increased above normal, hypotonic, increased above normal, hypotonic, decreased, increased, decreased

Acid-Base Balance (manual page 224)

A. 1. acids

2. base

3. 7.0, 7.4, 6.8, 8.0

4. chemical buffer system

5. [H_2CO_3], [HCO_3^-]

B. 1. T

2. T

3. F (The respiratory system is capable of compensating only up to 75% of the way toward normal.)

4. T

5. T

C. 1. a,d,e

2. b,e

3. a,b,c,

4. c

D. 1.a, 2.d, 3.c, 4.b

E. 1. metabolic acidosis

2. c

3. 7.1

4. respiratory alkalosis

5. d

6. 7.7

7. respiratory acidosis

8. a

9. 7.1

10. metabolic alkalosis

11. b

12. 7.7

F.

Lines of Defense	Speed of Action
Chemical buffer systems	Immediate (fractions of a second)
Respiratory control of pH	Few minutes
* Renal control of pH	Hours to days

Chapter 16 Digestive System

Introduction (manual page 230)

A. 1. F (The digestive system does not vary nutrient, water, or electrolyte uptake based on body needs, but instead it absorbs all digestible food that is ingested.)

2. T

3. F (Only when slow-wave potentials reach threshold do action potentials occur and resultant contractile activity takes place.)

B. 1. motility, secretion, digestion, absorption

2. excitatory, inhibitory

3. long, short

C.

	category of foodstuff	absorbable unit
1.	carbohydrates (polysaccharides, disaccharides)	monosaccharides (especially glucose)
2.	proteins	amino acids
3.	fats (triglycerides)	monoglycerides, free fatty acids

D. 1.c, 2.e, 3.a, 4.c, 5.g, 6.b, 7.h, 8.d, 9.c, 10.f

E. 1.b, 2.c, 3.d, 4.a, 5.b

Mouth; Pharynx and Esophagus (manual page 233)

A. 1. pharynx

2. swallowing center, intrinsic nerve plexus

B. 1. F (Sympathetic and parasympathetic stimulation both increase salivary secretion, but sympathetic stimulation produces a small volume of thick, mucus-rich saliva whereas parasympathetic stimulation produces a large volume of watery, enzyme-rich saliva.)

2. T

C. 1. e

2. b

D. 1.c, 2.e, 3.b, 4.a, 5.f, 6.d

Stomach (manual page 237)

A. 1. F (The most important function of the stomach is to store ingested food until it can be emptied into the small intestine at a rate appropriate for optimal digestion and absorption.)

2. F (Peristaltic waves only occur when the BER brings the smooth muscle to threshold, which depends on the muscle's level of excitability.)

3. T

4. F (Gastric secretion begins during the cephalic phase by means of a vagally-mediated reflex triggered by stimuli in the head, such as thinking about, smelling, or eating food.)

5. T

6. F (Individuals with gastric ulcers usually have a low gastric acid content, most likely because much of the acid has diffused into the damaged stomach wall.)

7. F (No absorption of foodstuffs occurs in the stomach. Absorption begins in the small intestine.)

B. 1. chyme

2. body, antrum

3. fat in the duodenum

4. secretin, cholecystokinin, gastric inhibitory peptide

5. protein in the stomach, acid and pepsinogen secretion, high H^+ concentration

6. body, salivary amylase, antrum, pepsin

C. 1.b, 2.d, 3.c, 4.a, 5.b

D. 1.c, 2.c, 3.d, 4.a, 5.e, 6.c, 7.a, 8.b, 9.c, 10.b, 11.d, 12.b

Pancreatic and Biliary Secretions (manual page 241)

A. 1.b, 2.d, 3.a, 4.c

B. 1. T

2. F (The major hormones secreted by the endocrine pancreas are insulin and glucagon. Secretin and CCK, which are secreted by the small intestine, act on the exocrine pancreas.)

3. F (The principal clinical manifestation of pancreatic insufficiency is impairment of fat digestion. Enzymes from non-pancreatic sources contribute to protein and carbohydrate digestion, but pancreatic amylase is the only enzyme for fat digestion.)

4. T

5. F (The liver secretes bile; the gall bladder stores and concentrates bile.)

6. T

C. 1. aqueous alkaline

2. duct, aqueous alkaline fluid, acinar, digestive enzymes

3. bile salts

4. bilirubin

5. jaundice

D. 1. d

2. a

3. b

Small Intestine (manual page 246)

A. 1. F (The main function of the ileocecal juncture is to prevent the bacterial-laden contents of the large intestine from contaminating the nutrient-rich small intestine.)

2. F (The protein components of digestive secretions and sloughed epithelial cells are digested and absorbed.)

3. F (Most of the fluid presented to the small intestine for absorption is from secreted digestive juices.)

4. T

B. 1. segmentation

2. circular folds, villi, microvilli

3. three

4. vitamin B_{12}, bile salts

5. calcium, iron

C. 1. e

2. a, b, c, e

3. a

Large Intestine (manual page 249)

A. 1. F (The large intestine does not have the ability to absorb nutrients.)

2. T

3. F (Symptoms associated with constipation are attributable to prolonged distention of the large intestine, particularly the rectum. Potentially toxic substances are removed by the liver before they reach the systemic circulation.)

4. F (To expel flatus, the external anal sphincter and abdominal muscles are both voluntarily contracted. Intestinal gas is forced out through the narrowed sphincter by the high intra-abdominal pressure. Narrowing of the sphincter prevents the escape of feces along with the flatus.)

B. 1. haustral, mass movements

2. an alkaline mucus solution

Chapter 17 Energy Balance and Temperature Regulation

Energy Balance (manual page 252)

A. 1. T

2. F (The excess energy is stored in the body, primarily as adipose tissue.)

3. F (Only about 75% of the chemical energy in nutrient molecules is harnessed to do biological work.)

4. T

5. F (Oxygen does not contain heat energy. For each liter of O_2 consumed in oxidizing nutrient molecules, 4.8 kilocalories of heat are liberated from the food on the average.)

6. T

7. T

8. F (On the average, fat people do not eat any more than thin people on a daily basis.)

B. 1. external

2. internal

3. metabolic rate

4. kilocalorie

5. basal metabolic rate (BMR)

6. specific dynamic action of food

7. food intake

8. hypothalamus

9. anorexia nervosa

C. 1. b,c

2. a

Temperature Regulation (manual page 257)

A. 1. F (Oral temperature is slightly lower than core temperature.)

2. F (Body temperature normally varies several degrees Fahrenheit.)

3. T

4. T

5. T

6. T

7. T

8. T

9. T

10. F (No sweating occurs during heat stroke, despite the fact that the body temperature is rapidly climbing, because of complete breakdown of the hypothalamic temperature-control centers.)

B. 1. humidity

2. hypothalamus

3. thermoreceptors

4. shivering

5. nonshivering thermogenesis

6. temperature gradient

7. sweating

8. heat stroke

9. frostbite

10. hyperthermia

C. 1. b

2. e

D. 1. a,c

2. a,d

3. b,d

4. b,c

5. b,c

6. b,d

7. a,d

8. a,d

9. b,d

10. b,c

11. a,c

12. a,d

E. 1.b, 2.a, 3.c, 4.a, 5.d, 6.c, 7.b, 8.d, 9.c, 10.b

Chapter 18 Principles of Endocrinology; Central Endocrine Organs

Introduction; General Principles of Endocrinology (manual page 264)

A. 1.c, 2.d, 3.b, 4.a

B. 1. target

2. tropic

3. liver

4. urinary excretion

5. down regulation

6. permissiveness

C. 1. T

2. T

3. F (Some endocrine glands exert non-endocrine effects in addition to secreting hormones.)

4. T

5. F (Target-tissue receptors are highly specific; they only bind with a particular hormone.)

6. T

7. T

8. F (Each steroidogenic organ has a limited set of enzymes for producing only a given type or types of steroid hormones.)

9. T

10. F (Metabolic modification of some hormones results in their activation.)

11. T

D. 1. c

2. b

E.

1. a	11. h	21. f
2. g	12. d	22. f
3. b	13. e	23. e
4. g	14. f	24. j
5. e	15. e	25. j
6. f	16. b	26. j
7. a	17. d	27. d
8. a	18. b	28. c
9. e	19. j	29. b
10. b	20. e	30. a

Hypothalamus and Pituitary (manual page 271)

A. 1. b
2. a
3. b
4. a
5. b
6. b
7. a
8. c
9. a
10. b
11. a
12. b
13. b
14. a
15. b
16. b

B. 1. e
2. a
3. g
4. d
5. g
6. a
7. f
8. b
9. e
10. g
11. c
12. h
13. b
14. g
15. e
16. f

C. 1. F (Melanocyte-stimulating hormones regulate skin coloration in certain lower species but do not control skin color in humans.)
2. T
3. T
4. F (It is long-loop negative feedback.)

D. 1. c,e
2. b

E. 1.c, 2.b, 3.b, 4.a, 5.a, 6.c, 7.c

Hormonal Control of Growth (manual page 276)

A. 1. T

2. T

3. T

4. T

5. T

B. 1. osteoblasts

2. osteoclasts

3. osteocytes

4. epiphyseal plate

5. somatomedins

C. 4

D. 1, 2, 3, 5, 6, 7, 9, 10, 11, 12, 13, 15, 17, 18

Chapter 19 Peripheral Endocrine Organs

Thyroid Gland (manual page 282)

A. 1. F (The follicular cell engulfs a portion of colloid by endocytosis and frees thyroid hormone from thyroglobulin. The lipophilic thyroid hormone then diffuses through the plasma membrane of the follicular cell to enter the blood.)

2. T

3. F (MIT and DIT remain in the follicular cell and are deiodinated. The freed iodine is recycled for synthesis of more hormone.)

4. T

5. T

6. T

7. T

8. F (Only after a delay of several hours is the metabolic response to thyroid hormone first detectable, and the maximal response is not evident for several days.)

9. T

10. T

11. T

12. F (A goiter may also occur in association with thyroid hyposecretion, as is the case with iodine deficiency or thyroid gland malfunction.)

B. 1. follicular, T_4 (thyroxine or tetraiodothyronine), T_3 (triiodothyronine), C, calcitonin

2. T_4, T_3

3. colloid, thyroglobulin

4. tyrosine

5. iodide pump, iodide trapping mechanism

6. MIT (monoiodotyrosine), DIT (diiodotyrosine)

7. T_4, T_3

8. T_4, T_3

9. thyroxine-binding globulin, albumin, thyroxine-binding prealbumin

10. TSH (thyroid-stimulating hormone)

11. goiter

12. cretinism

C. 1. c,e

2. d

3. c

Adrenal Gland and Stress (manual page 288)

A. 1. adrenal cortex, steroids

2. adrenal medulla, catecholamines

3. pro-opiomelanocortin

4. stress

B. 1. T

2. F (Aldosterone is primarily bound to albumin. Cortisol is bound to corticosteroid-binding globulin.)

3. T

4. T

5. F (Hyperphosphatemia is not present. Symptoms include hypernatremia, hypokalemia, and hypertension.)

6. F (Epinephrine and norepinephrine have differing affinities for different adrenergic receptor types and consequently exert some differing effects.)

7. T

C. 1. c

2. b,d

3. d

4. c

5. b

6. a,b

7. c

8. b

9. a,b,c

10. a

11. b,c,d

12. a

13. a

14. d

15. c

16. c

17. b

18. a

19. b

20. c

21. a,b

22. b

23. a,b

24. b

25. b,d

26. c

27. b

28. d

29. a

30. b,c (and d)

31. a

32. a

33. b

34. a

D. 1.

Categories of Hormones	Principal Hormone in Category
a. mineralocorticoid	aldosterone
b. glucocorticoid	cortisol
c. sex hormones	dehydroepiandrosterone

2. increased epinephrine, increased CRH-ACTH-cortisol, increased glucagon, decreased insulin, increased renin-angiotensin-aldosterone, increased vasopressin

Endocrine Control of Fuel Metabolism (manual page 293)

A. 1. F (Structural proteins within cells are continuously being degraded and replaced.)

2. T

3. T

4. F (Type II diabetes mellitus is caused by a reduced sensitivity of the target cells to insulin, not by a lack of insulin secretion.)

5. T

6. F (Type II diabetics do not require insulin injections.)

7. T

8. T

9. T

B. 1. intermediary metabolism or fuel metabolism

2. anabolism

3. catabolism

4. brain

5. glycogenesis

6. glycogenolysis

7. gluconeogenesis

8. liver

9. ketone bodies

10. brain, working muscles, liver

C. 1. e

2. c

D. 1. glucose 2. glycogen

3. free fatty acids 4. triglycerides

5. amino acids 6. body proteins

E. 1. a, b,c, g, i, j

2. b,c,d,f,g,h,i,j,l

3. a,b,c,d,f,g,h,i,j,k.l,m,n,o,q,r,s

Endocrine Control of Calcium Metabolism (manual page 299)

A. 1. T

2. F (The most life-threatening consequence of hypocalcemia is overexcitability of nerves and muscles.)

3. F (Absorption of ingested Ca^{++} from the intestine is controlled by vitamin D.)

4. T

5. F (The labile pool of Ca^{++} is in the bone fluid. The stable pool of Ca^{++} is in the bone crystals.)

6. T

7. F (Hyperparathyroidism is characterized by hypercalcemia and hypophosphatemia.)

B. 1.c; 2.a; 3. a,b; 4.c; 5.a; 6.c; 7.b

C. 1.b; 2.a; 3.a,b; 4.b

D. 1. freely diffusible

2. parathyroid hormone (PTH), calcitonin, vitamin D

3. bone, kidneys, digestive tract

4. hydroxyapatite

5. C cells, thyroid gland

6. skin

7. liver, kidneys

8. kidneys, PTH, plasma phosphate

E. 1. a, c, d, e, g

2. d, f

Chapter 20 Reproductive Physiology

Introduction (manual page 305)

A. 1. gonads, testes, ovaries

2. gametes, ova (eggs), spermatozoa (sperm), estrogen, progesterone, testosterone

3. accessory sex glands

4. external genitalia

5. secondary sexual characteristics, testosterone, estrogen

B. 1. T

2. F (Testosterone is converted into dihydrotestosterone, which promotes the development of the undifferentiated external genitalia alone male lines.)

3. T

C. 1.c, 2.b, 3.a, 4.e, 5.a, 6.b, 7.c, 8.d

D. 1. c

2. b

Male Reproductive Physiology; Sexual Intercourse between Males and Females (manual page 310)

A. 1. T

2. F (Spermatogonia line the outer perimeter of the seminiferous tubules; spermatozoa are found at the lumen.)

3. T

4. T

5. T

6. F (Secretory bursts of GnRH occur once every two to three hours.)

7. F (At puberty, the hypothalamus becomes less sensitive to feedback inhibition by testosterone.)

8. F (Melatonin secretion decreases during exposure to light.)

9. T

10. F (Erection is accomplished by engorgement of the vascular columns of erectile tissue with blood.)

11. F (The clitoris becomes erect during sexual arousal.)

12. T

13. T

14. F (The bladder sphincter is closed to prevent sperm from entering the bladder and urine from leaving the bladder.)

15. T

16. F (The outer third of the vagina narrows and tightens around the penis whereas the upper two thirds of the vagina expands to create a space for ejaculate deposition.)

17. T

B. 1. seminiferous tubules, FSH, testosterone

2. Leydig cells, LH

3. cryptorchidism

4. head, acrosome, midpiece, tail

5. excitement, plateau, orgasmic, resolution

6. ejaculation, emission, expulsion

C. 1.c, 2.a, 3.b, 4.c, 5.a, 6.a, 7.c, 8.b, 9.e

D. 1.a, 2.b, 3.c, 4.d, 5.d

E. 1.c, 2.f

Female Reproductive Physiology (manual page 318)

A. 1. T

2. F (A follicle that fails to reach maturity undergoes atresia; that is, degenerates and forms scar tissue.)

3. T

4. F (The oviduct is the site of fertilization.)

5. T

6. F (The second stage involves birth of the baby. The third stage involves delivery of the placenta.)

B. 1. LH

2. corpus luteum

3. thecal, LH, granulosa, FSH

4. fetal adrenal cortex, placenta

5. inner cell mass

6. corpus luteum of pregnancy

7. human chorionic gonadotropin

C. 1. a

2. c

3. b

4. a

5. a

6. b

D. 1. a

2. a

3. b

4. b

5. b

6. a

7. a

8. c

E. 1. d

2. e

3. d

4. a

5. d

APPENDIX B: SUPPLEMENTAL READING LIST

The following list includes textbooks at a more advanced level and relevant articles in lay science journals. The list is not meant to be comprehensive but merely as a starting point for those interested in pursuing specific topics in more depth. Publications in specialty science journals are not included.

General Physiology References

Berne, R.M. and M.N. Levy (eds). Physiology, Second edition, C.V. Mosby Company, St. Louis, 1988

Guyton, A.C. Textbook of Medical Physiology, Seventh edition, W.B. Saunders Company, Philadelphia, 1986

Mountcastle, V.B. (ed). Medical Physiology, Fourteenth edition, (two volume set), C.V Mosby Company, St. Louis, Vol. I 1980, Volume II 1979

Schmidt, R.F. and G. Thews (eds). Human Physiology, Springer-Verlag, Berlin, 1983

Vick, R.L. Contemporary Medical Physiology, Addison-Wesley Publishing Company, Inc., Menlo Park, California, 1984

General Anatomy References

Bacon, R.L. and N.R. Niles. Medical Histology: A Text-Atlas with Introductory Pathology, Springer-Verlag, New York, 1983

Bloom, W. and D.W. Fawcett. A Textbook of Histology, Eleventh edition, W.B. Saunders Company, Philadelphia, 1986

Cormack, D.H. Introduction to Histology, J.B. Lippincott Company, Philadelphia, 1984

Ham, A.W. Histology, Ninth edition, J.B. Lippincott Company, Philadelphia, 1987

Kessel, R.G. and R.H. Kardon. Tissues and Organs: A Text-Atlas of Scanning Electron Microscopy, W.H. Freeman and Company, San Francisco, 1979

Spence, A.P. Basic Human Anatomy, Second edition, Benjamin-Cummings Publishing Company, Menlo Park, California, 1986

Tortora, G.J. Principles of Human Anatomy, Fourth edition, Harper and Row, Publishers, New York, 1986

Warwick, R. and P. Williams, Gray's Anatomy, Thirty-fifth edition, W.B. Saunders, Philadelphia, 1973

Woodburne, R.T. and W.E. Burkel. Essentials of Human Anatomy, Eighth edition, Oxford University Press, New York, 1988

Exercise Physiology References

Brooks, G.A. and T.D. Fahey, Exercise Physiology: Human Bioenergetics and Its Applications, John Wiley and Sons, New York, 1984

DeVries, H.A. Physiology of Exercise, Fourth edition, W.C. Brown, Dubuque, Iowa, 1986

Lamb, D.R. Physiology of Exercise: Responses and Adaptations, MacMillan Publishing Company, New York, 1984

McArdle, W.D., F.D. Katch, and V.L. Katch. Exercise Physiology: Energy, Nutrition, and Human Performance, Second edition, Lea and Febiger, Philadelphia 1986

Principles of Homeostasis and Regulation (Chapter 1)

Adolph, E.F. (ed). The Development of Homeostasis, Academic Press, Inc., New York, 1960

Adolph, E.F. (ed). Origins of Physiological Regulations, Academic Press, Inc., New York, 1968

Jones, R.W. Principles of Biological Regulation: An Introduction to Feedback Systems, Academic Press, Inc., New York, 1973

Cellular Physiology (Chapters 2 and 3)

Alberts, B., D. Bray, J. Lewis, M. Raff, K. Roberts, J. Watson, Molecular Biology of the Cell, Second edition, Garland Publishing, Inc., New York, 1989

Allen, R.D. "The Microtubule as an Intracellular Engine," Scientific American, Volume 256 Number 2, February 1987

Berridge, M.J. "The Molecular Basis of Communication within the Cell," _Scientific American_, Volume 253 Number 4, October 1985

Bretscher, M.S. "The Molecules of the Cell Membrane," _Scientific American_, Volume 253 Number 4, October 1985

Byrne, J.H. and S.G. Schultz. _An Introduction to Membrane Transport and Bioelectricity_, Raven Press, New York, 1988

Carafoli, E. and J.T. Penniston, "The Calcium Signal," _Scientific American_, Volume 253 Number 5, November 1985

Darnell, J., H. Lodish, and D. Baltimore, _Molecular Cell Biology_, Scientific American Books, New York, 1986

The Molecules of Life: Readings from Scientific American, Scientific American Books, New York, 1986

Ptashne, M. "How Gene Activation Works," _Scientific American_, Volume 260 Number 1, January 1989

Radman, M. and R. Wagner. "The High Fidelity of DNA Duplication," _Scientific American_, Volume 259 Number 2, August 1988

Rothman, J.E. "The Compartmental Organization of the Golgi Apparatus," _Scientific American_, Volume 253 Number 3, September 1985

Stahl, F.W. "Genetic Recombination," _Scientific American_, Volume 256 Number 2, February 1987

Weber, K. and M. Osborn. "The Molecules of the Cell Matrix," _Scientific American_, Volume 253 Number 4, October 1985

Nervous System (Chapters 4, 5, 6, and 7)

Aoki, C. and P. Siekevitz. "Plasticity in Brain Development," _Scientific American_, Volume 259 Number 6, December 1988

Bradford, H.F. _Chemical Neurobiology: An Introduction to Neurochemistry_, Scientific American Books, New York, 1986

Fitzgerald, M.J.T. _Neuroanatomy Basic and Applied_, Bailliere Tindall, London, 1985

Goldstein, G.W. and A.L. Betz. "The Blood-Brain Barrier," _Scientific American_, Volume 255 Number 3 September 86

Kandel, E.R. and J.H. Schwartz. Principles of Neural Science, Elsevier, New York, 1985

Kuffler, S.W., J.G. Nicholls, and A.R. Martin. From Neuron to Brain, Second edition, Sinauer Associates, Inc., Publishers, Sunderland, Mass., 1984

Krut, Z.L. and C.J. Pycock. Neurotransmitters and Drugs, Second edition, University Park Press, Baltimore, 1983

Livingstone, M. "Art, Illusion, and the Visual System," Scientific American, Volume 258 Number 1, January 1988

Masland, R.H. "The Functional Architecture of the Retina," Scientific American, Volume 255 Number 6, December 1986

McKean, K. "Pain", Discover, October 1986

Nathans, J. "The Genes for Color Vision," Scientific American, Volume 260 Number 2, February 1989

Ottoson, D. Physiology of the Nervous System, Oxford University Press, New York, 1983

Petit, T.L. and G.O. Ivy (eds). Neural Plasticity: A Lifespan Approach, Alan R. Liss, Inc., New York, 1987

Schnapf, J.L. and D.A. Baylor. "How Photoreceptor Cells Respond to Light," Scientific American, Volume 256 Number 4, April 1987

Springer, S.P. and G. Deutsch. Left Brain, Right Brain, Scientific American Books, New York, 1985

Stryer, L. "The Molecules of Visual Excitation," Scientific American, Volume 251 Number 1, July 1987

Wang, M. and A. Freeman, Neural Function, Little, Brown and Company, Boston, 1987

Muscle (Chapter 8)

Bourne, G.H. (ed). The Structure and Function of Muscle, Second edition, (four volumes), Academic Press, New York, Vol. I 1972, Vols. II and II 1973, Vol. IV 1974

Entman, M.L., and W.B. VanWinkle. Sarcoplasmic Reticulum in Muscle Physiology, Boca Raton, Florida, CRC Press, 1986

Gaesser, G.A. and G.A. Brooks. "Metabolic bases of excess post-exercise oxygen consumption: a review," Medicine and Science in Sports and Exercise, Volume 16 Number 1, 1984

Huxley, A. Reflections on Muscle, Princeton University Press, Princeton, New Jersey, 1980

Junge, D. Nerve and Muscle Excitation, Sinauer Associates, Sunderland, Mass., 1980

Keynes, R.D. and D.J. Aidley. Nerve and Muscle, Cambridge University Press, Cambridge, 1981

Peachy, L.D., R.H. Adrian, S.R. Geiger. Skeletal Muscle, Williams and Wilkins, Baltimore, 1983

Circulatory System (Chapters 9, 10, and 11)

Brown, M.S. and J. Goldstein, "How LDL Receptors Influence Cholesterol and Atherosclerosis," Scientific American, Volume 251 Number 5, November 1984

Goerke, J. and A. H. Mines. Cardiovascular Physiology, Raven Press, New York, 1988

Little, R.C. Physiology of the Heart and Circulation, Year Book Medical Publishers, Inc., Chicago, 1985

Mohrman, David E. and L.J. Heller. Cardiovascular Physiology, McGraw Hill Book Company, New York 1986

Robinson, T.F., S.M. Factor, and E.H. Sonnenblick. "The Heart as a Suction Pump," Scientific American, Volume 254 Number 6, June 1986

Shepherd, J.T. and P.M. Vanhoutte, The Human Cardiovascular System: Facts and Concepts, Raven Press, New York 1979

Smith, J.J. and J.P. Kampine, Circulatory Physiology, Williams and Wilkins, Baltimore, 1984

Immune Defense (Chapter 12)

Barrett. J.T. Textbook of Immunology, Fourth edition, Mosby, St. Louis, 1983

Cohen, I.R. "The Self, the World, and Autoimmunity," Scientific American, Volume 258 Number 4, April 1988

Edelson, R.L. and J.M. Fink. "The Immunologic Function of Skin," Scientific American, Volume 252 Number 6, June 1985

Greaves, M.F., J.J.T. Owen, and M.C. Raff. "T and B Lymphocytes: Origins, Properties, and Roles in Immune Responses, Excerpta Medica, Amsterdam, 1973

Marrack, P. and J. Kappler, "The T Cell and Its Receptor", Scientific American, Volume 254 Number 2, February 1986

Milstein, C. "From Antibody Structure to Immunological Diversification of Immune Response", Science, Volume 231, March 1986

Tonegawa, S. "The Molecules of the Immune System," Scientific American, Volume 253 Number 4, October 1985

Respiratory System (Chapter 13)

Mines, A.H. Respiratory Physiology, Second edition, Raven Press, New York, 1986

Taylor, A.E., K. Rehder, R.E. Hyatt, and J.C. Parker, Clinical Respiratory Physiology, W.B. Saunders, Philadelphia, 1989

West, J.B. Respiratory Physiology: The Essentials, Third edition, Williams and Wilkins, Baltimore, 1985

Urinary System, Body Fluids, Acid-Base Balance (Chapters 14 and 15)

Cantin, M. and J. Genest. "The Heart as an Endocrine Gland," Scientific American, Volume 254, Number 2, February 1986

Davenport, H.W., The ABC of Acid-Base Chemistry, Seventh edition, University of Chicago Press, Chicago, 1978

Gottschalk, C.W., R.W. Berliner, and G.H. Giebisch. Renal Physiology: People and Ideas, Oxford University Press, New York, 1987

Haperin, M.L. and M.B. Goldstein, Fluid, Electrolyte and Acid-Base Emergencies, W.B. Saunders Company, Philadelphia, 1988

Marsh, D.J. Renal Physiology, Raven Press, New York, 1983

Klalr, S. The Kidney and Body Fluids in Health and Disease, Plenum Medical Book Company, New York, 1984

Sullivan, L.P., and J.J. Grantham. Physiology of the Kidney, Second edition, Lea and Febiger, Philadelphia, 1982

Valtin, H. Renal Function: Mechanisms Preserving Fluid and Solute Balance in Health, Little, Brown and Company, Boston, 1983

Vander, A.J. Renal Physiology, Third edition, McGraw Hill Book Company, New York, 1985

Digestive System (Chapter 16)

Davenport, H.W. A Digest of Digestion, Second edition, Year Book Medical Publishers, Inc., Chicago, 1978

Davenport, H.W. Physiology of the Digestive Tract, Fifth edition, Year Book Medical Publishers, Inc., Chicago, 1982

Granger, D.N. J.A. Barrowman, and D.R. Kvietys. Clinical Gastrointestinal Physiology, W.B. Saunders Company, Philadelphia, 1985

Johnson, L.R., J. Christensen, M.I. Grossman, E.D. Christensen, S.G. Schultz (eds). Physiology of the Gastrointestinal Tract, (two volumes), Raven Press, New York, 1981

Sernka, T. and E. Jacobson, Gastrointestinal Physiology: The Essentials, Williams and Wilkins Company, Baltimore, 1983

Energy Balance and Temperature Regulation (Chapter 17)

Bligh, J. Temperature Regulation in Mammals and Other Vertebrates, American Elsevier Publishing Company, New York, 1973

Lytle, L.D. "Control of Eating Behavior", Nutrition and the Brain, R.J. Wurtman and J.J. Wurtman (eds), Raven Press, New York, 1977

Also see Mountcastle, general physiology reference

Endocrine and Reproductive Systems (Chapters 18, 19, and 20)

Burger,H. and D. DeKretser. Comprehensive Endocrinology: The Testis, Raven Press, New York, 1981

Carmichael, S.W. and H. Winkler, "The Adrenal Chromaffin Cell," Scientific American, Volume 253 Number 2, August 1985

Crapo, L. Hormones: Messengers of Life, Scientific American Press, New York, New York, 1985

Hedge, G.A., H.D. Colby, and R.L. Goodman. Clinical Endocrine Physiology, W.B. Saunders Company, Philadelphia, 1987

Martin, C.R. Endocrine Physiology, Oxford University Press, New York, 1985

Orci, L., J. Vassali, and A. Perrelet, "The Insulin Factory", Scientific American, Volume 259 Number 3, September 1988

Serra, G. Comprehensive Endocrinology: The Ovary, Raven Press, New York, 1983

Snyder, S.H. "The Molecular Basis of Communication between Cells," Scientific American, Volume 253 Number 4, October 1985

Tepperman, J. Metabolic and Endocrine Physiology, Fifth edition, Year Book Medical Publishers, Inc. Chicago, 1987